THE
AMERICAN
WAY
OF
IRREGULAR
WAR

AN ANALYTICAL MEMOIR

CHARLES T. CLEVELAND
WITH DANIEL EGEL

T0146135

RAND NATIONAL SECURITY RESEARCH DIVISION

July 2020

For more information on this publication, visit www.rand.org/t/PEA301-1

Library of Congress Cataloging-in-Publication Data is available for this publication.
ISBN: 978-1-9974-0544-9

Support RAND

Make a tax-deductible charitable contribution at
www.rand.org/giving/contribute

www.rand.org

Preface

American irregular warfare is the United States' unique, and in recent times troubled, approach to conflict in which armed civilian or paramilitary forces, and not regular armies, are the primary combatants. In most forms, it emphasizes the importance of local partnerships and gaining legitimacy and influence among targeted populations. It is thus a critical capability in contests where populations, rather than territory, are decisive. This analytical memoir draws on my nearly four decades of experience in the U.S. Army to explore the strengths and limitations of America's current irregular warfare capability and provide recommendations for what the United States must do to develop the world-class American way of irregular war it needs.

I wrote this memoir because I remain deeply troubled by the fact that the United States has failed to achieve its strategic objectives in nearly every military campaign in which I was involved. Time and time again I saw tremendous success at the tactical level, whether it was the war on drugs in Bolivia, peace enforcement in Bosnia, counterinsurgencies in Afghanistan and Iraq, or partnered counterterrorism operations in Lebanon and Yemen. But in each case, American tactical brilliance was followed by strategic muddling and eventual failure. The United States failed, in large part, because it has not developed the understanding, capabilities, and structures necessary to achieve strategic successes in population-centric contests, which are won and lost by controlling and influencing populations rather than occupying territory.

In the wake of the coordination failure that led to the failed Operation Eagle Claw and the intelligence failure that led to September 11, it took action by Congress and the support of the President to drive the reforms that we needed. I believe that Congress and the President will need to act again. To provide a proactive defense against the irregular warfare campaigns of our enemies and the necessary offensive potential to destabilize our Great Power adversaries, we must turn to, and not away from, the American way of irregular war.

This research was sponsored by the Smith Richardson Foundation and conducted within the International Security and Defense Policy Center of the RAND National Security Research Division (NSRD). NSRD conducts research and analysis for the Office of the Secretary of Defense, the U.S. Intelligence Community, U.S. State Department, allied foreign governments, and foundations.

For more information on the RAND International Security and Defense Policy Center, see www.rand.org/nsrd/isdp or contact the director (contact information is provided on the webpage).

To Mary Ann, for her encouragement, patience, understanding, and love.

To my Dad, Walter, for the right start and support along the way.

To Dobromir Neikov, Kenny McMullin, Roy Trumble, and Geoff Lambert, for their inspiration and leadership.

To the men and women who fight America's irregular wars, for their sacrifice and service.

Contents

Preface iii

Figures xi

Summary xiii

Acknowledgments xxv

Abbreviations xxvii

CHAPTER ONE

Introduction 1

An American Way of Irregular War 3

Prevalence of Population-Centric Conflict 7

America's Irregular Warfare Misadventures 9

Analytical Approach 13

Organization of the Analytical Memoir 15

CHAPTER TWO

Irregular Warfare and the Cold War in Europe 19

Unconventional Warfare in Europe: 10th Special Forces in the Cold
    War 20

Emergence of Strategic Reconnaissance: The Beginning of Special
    Forces' Transition to a Supporting Role 23

Operation Eagle Claw: The American Pivot Toward Counterterrorism ... 26

CHAPTER THREE

**Bolivia and the Department of Defense's Entry into the War on
    Drugs** 29

Building Bolivia's Counternarcotics Capability: Success at the Tactical
    Level 32

Unprepared for This New Form of Population-Centric Warfare:
    Difficulties Above the Tactical Level 37
Strategic Failure in the U.S. Mission in Bolivia: Inadequacy of Existing
    Models and Structures for This Population-Centric Conflict 41

CHAPTER FOUR
El Salvador and the Fight Against Communism in the Americas 45
Building an El Salvadoran Counterinsurgency Capability: Provisional
    Success at the Tactical Level 48
Professionalizing the Armed Forces of El Salvador: Success at the
    Operational Level 52
Strategic Success by Accident: Inadequacy of Existing Models and
    Structures for This Population-Centric Conflict 54

CHAPTER FIVE
Panama and the Transition from Traditional to Irregular 59
Planning for Regular and Irregular War: My First Special Operations
    Campaign Plan 62
Speed, Surprise, Overwhelming Firepower . . . and Good Preparation:
    Success in the "Traditional" Warfare Component of Just Cause 66
Postinvasion Unconventional Warfare: The Peaceful Surrender of
    Panamanian Security Forces 69
Securing the Peace: Transition into an Irregular Warfare Mission 73
Winning the Peace: More Luck Than Planning in the U.S. Strategic
    Success 76

CHAPTER SIX
The Decade of Delusion and My Pentagon Wars 81
Operation Desert Storm: Validation of America's New Conventional
    and Raiding Capabilities 84
My Pentagon Wars: Fighting for America's Irregular Warfare
    Capability 86
Developing Unconventional Warfare Campaigns: Creating Special
    Operations Policy Options 91

CHAPTER SEVEN
Peacekeeping in Bosnia and the Reemergence of Irregular Warfare ... 95
Rebirth of Unconventional Warfare: The Joint Commission Observer
    Mission 98

Communicating the Value of Irregular Warfare: Growing Pains in
    Achieving Unity of Effort 104
Organizing for Irregular Warfare: Growing Pains in Establishing the
    New Combined Joint Special Operations Task Force 108

CHAPTER EIGHT
Unconventional Warfare in the War on Terror 113
Building a Tactical Unconventional Warfare Capability for the 21st
    Century: 10th Special Forces Jedburgh Teams 115
Task Force Viking and Unconventional Warfare in Northern Iraq 118
Inadequate Irregular Warfare Capabilities Above the Tactical Level:
    Difficulties in Achieving Enduring Successes 128

CHAPTER NINE
Special Operations Campaigning in Latin America 131
Success Through Partnering: Operation Jaque 134
Creating Indigenous-Centric Options: Establishing a Special
    Operations Campaign Plan 138
Resourcing Irregular Warfare: The Derivative Benefit of the Special
    Operations Campaign 142
Relationship Building as a Mission-Essential Task: The First Special
    Operations Command Forward 145

CHAPTER TEN
At the Vanguard of American Irregular Warfare 149
Irregular Warfare Campaigns in Lebanon, Pakistan, and Yemen:
    Adapting Lessons from SOCSOUTH 151
Designing and Executing Irregular Warfare Campaigns for the "Big"
    Wars: The Value of Operational-Level Special Operations
    Headquarters 156
Organizing for Population-Centric Conflict: The Human Domain 164

CHAPTER ELEVEN
Shepherding America's Irregular Warfare Capability 169
Developing Doctrine and a Ten-Year Vision for Army Special
    Operations: A Foundation for Reforming America's Approach
    to Irregular Warfare 172
Redesigning USASOC's Irregular Warfare Capabilities:
    Sequestration as a Forcing Function 179

The Beginnings of an American Way of Irregular War: Department
    of Defense–Wide Efforts to Enhance America's Irregular
    Warfare Capability.. 186
An American Way of Irregular War: My Postscript 192

CHAPTER TWELVE
Key Observations 195
Observation 1: U.S. Tactical-Level Formations Have Performed
    Admirably in Irregular Warfare Campaigns........................... 196
Observation 2: Irregular Warfare Missions Require Irregular Warfare
    Campaigns.. 198
Observation 3: The U.S. Military Is Not Well Organized for
    Irregular Warfare Campaigns 201
Observation 4: The United States Lacks the Concepts, Doctrine, and
    Canon Necessary to Be Effective in Population-Centric
    Conflicts 203
Observation 5: There Is Insufficient Professional Military Education
    for Irregular Warfare.. 207
Observation 6: The U.S. National Security Enterprise Is Structured
    to Fail in Population-Centric Conflicts.............................. 210

CHAPTER THIRTEEN
Recommendations.. 213
Recommendation to Congress: Mandate an Independent Review of
    U.S. Strategic Failure in Population-Centric Conflicts 215
Recommendation to the President: Reorganize the Executive Branch
    Around the Security Challenges of the 21st Century.................. 218
Recommendation to Concerned Citizens: Establish an Institution
    Outside Government Dedicated to Understanding American
    Irregular Warfare 221

CHAPTER FOURTEEN
Conclusion 223

References 227

Figures

1.1. Prevalence of Population-Centric Warfare Since World
      War II 8
1.2. U.S. Involvement in Population-Centric Conflicts
      Post–World War II... 10
11.1. Core Competencies of Army Special Operations
      (the "Cardona Slide").. 177

Summary

My reason for writing this analytical memoir is simple, though deeply unsettling: During my career, which spanned from 1978 to 2015, the United States failed to achieve its strategic objectives in nearly every military campaign in which I was involved. Time and time again I saw tremendous success at the tactical level (never once was a U.S. tactical-level formation defeated) followed by strategic muddling and eventual failure. The costs of these strategic failures, in wasted life and treasure, have been far too high.

We have consistently failed in these population-centric conflicts because we have chosen not to mature our irregular warfare capabilities. Irregular warfare has been a central component of American warfare throughout our history, but interest in and resourcing for this form of warfare have waxed and waned and increasingly failed to keep up with changes in society that have made irregular war today's most vexing form of conflict. In the wake of the great wars, the United States built and has sustained a world-class capability for traditional and nuclear warfare. Yet we have not done the same for irregular warfare.

My experience, world events since the Vietnam War, and the writings of our Great Power adversaries lead me to believe that irregular warfare will remain the most prevalent form of conflict facing the United States for many years to come. Although deterring the aggression of our adversaries will certainly require that the United States expand the lethality of its conventional and nuclear capabilities, to sustain its military supremacy, actual competition and conflict with these nation-states will most likely be irregular. This reflects a deliberate calculus by our adversaries, who understand that conventional or nuclear war with the United States is risky, costly, and ultimately unnecessary. These adversaries have learned all too well that they can impose tremendous costs on the United States using irregular approaches.

Maturing the American way of irregular war by building the understanding, capabilities, and structures necessary to achieve strategic successes in these conflicts is thus an imperative. This study provides recommendations from a practitioner's perspective for how the United States might do so. I admit I am far from America's most combat-proven irregular warfare expert, nor am I its most artful and learned herald—my analysis is therefore likely both imperfect and incomplete. But my experience and a deepening regret for the losses inflicted on our best young men and women have compelled me to say something, in hopes that it might drive the change that I believe the United States needs to protect our way of life.

Context

Population-centric conflicts are irregular warfare contests that are won and lost by controlling and influencing populations rather than occupying territory. These contests have proven timeless and ubiquitous, and U.S. interests at home and abroad have been long threatened by civil war, insurgency, terrorism, resistance, and subversion. Revolutionaries and guerrillas, motivated by religion or ideology, are the foot soldiers in these conflicts; state sponsorship is covert; and terror is a preferred tactic. Advances in technology now allow the unassimilated, disaffected, and even hostile diaspora from across the globe to connect in new ways, virtually erasing national borders.

The United States has struggled in these contests. During the nearly four decades of my career, which began in the shadow of our failure in Vietnam, U.S. combat experiences revealed stunning military capabilities and repeated tactical successes yet an inability to successfully conclude contests in which our enemy could operate and shelter among indigenous people. Neither America's current irregular warfare capability, which has extremely limited ability above the tactical level, nor its more mature conventional forces have proven effective in these contests.

Despite this persistent difficulty, the Department of Defense has resisted undertaking a comprehensive investigation of the cause of these evident failures. The U.S. response to the lessons of Vietnam was to vow to never fight such conflicts again. As a result, the United States remained ill prepared for the population-centric contests in which I was involved during my career, and the military options offered to political leadership were frequently unsuitable and the warfighting concepts poorly matched for the adversaries we faced.

The United States, in my view, has three options at this point. The first is to once again declare, as we did after Vietnam, that we will never again fight in irregular wars and that a deterrent force will suffice to keep our adversaries in check. The second is to believe, as we have essentially done over the past two decades, that the U.S. force created for Great Power conflict is also sufficient for irregular warfare. The third is for the United States to elevate and mature its irregular warfare capability to a world-class level, comparable to its nuclear and conventional capabilities. It is my assessment that only this last approach would give the United States the expertise to succeed in irregular war—and also give policymakers the advice they need to be successful in this form of warfare.

About This Study

This analytical memoir is an exploration of the American way of irregular war. In the same way that today's conventionally focused American way of war is defined by America's technical capacity and technological edge, the American way of irregular war must be tied to our notions of religious pluralism, democracy, and, above all, human rights. The American way of war secures us from existential threats of near-peer powers and guarantees the lanes of global commerce, while the American way of irregular war protects our way of life by both promoting our worldview and giving people the tools to realize the same opportunities that we have had.

Core to this American way of irregular war is the ability to influence our allies, adversaries, and their populations. It requires us to work with and through local partners throughout the world—sometimes with those who do not fully share our values—to influence them and their societies. It is about the judicious application of force, when necessary, but also about creating mutual understanding and sharing the great opportunities that our great democracy has provided us. It requires a deep-enough understanding of populations to identify the right types of outcomes and the strategic patience necessary to achieve them.

This analysis relies heavily on my own personal experiences as a practitioner of American irregular warfare, which included more than a dozen population-centric contests. From humble beginnings as an intelligence officer in support of America's unconventional warfare efforts in Europe in the late 1970s, my service would be dominated by my time in Special Forces with direct involvement in our campaigns in Latin America (e.g., Bolivia, Colombia, El Salvador), Western Europe (e.g., Bosnia, Kosovo, Georgia), and the Middle East (e.g., Afghanistan, Iraq, Lebanon, Pakistan, Yemen). In this career, I experienced the strengths and weaknesses of America's irregular warfare capability. And during my time as the commander of U.S. Army Special Operations Command, where I closed out my career, I also sought to effect some modest improvements.

But this study is neither my memoir nor my biography. It traces the significant moments in my career in which I was witness to a part of America's irregular warfare capability, describing what I learned along the way. It then validates the insights and observations that emerge from this approach using a blend of primary and secondary sources. The primary data collection included consultations with more than 40 senior-level experts—former uniformed practitioners, intelligence officers, career diplomats, development professionals, and policy experts. Although it was simply impossible to provide a comprehensive review of the extremely broad and diverse literature on irregular warfare and America's approach to irregular warfare, this analysis also reviewed and references what I believe to be seminal texts for understanding American irregular warfare and research relevant to the specific irregular warfare efforts in which I was involved.

The result is a story of the American way of irregular war, one focused on U.S. Army Special Forces, typically the vanguard in America's irregular warfare efforts, but a story that includes the many other military and civilian elements that are Special Forces' indispensable partners. It is a story of the waxing and waning of interest and support for irregular warfare and the vulnerability that this lack of commitment created. It is a story of strategic successes and failures that have their root, at least in part, in the strengths and weaknesses of that irregular warfare capability. It is a story of what the American way of irregular war might look like and how that would enable us to contest the threats that we face today.

Key Observations

Six key overarching observations emerge from my time in the service and the four-year research process that led to this manuscript:

Observation 1: U.S. tactical-level formations have performed admirably in irregular warfare campaigns. Throughout my career, our tactical formations succeeded in every irregular warfare mission that I saw put before them. The intelligence, creativity, flexibility, and professionalism of operators across our national security enterprise are the reason we never lost a battle. This was certainly true of Special Forces, the only U.S. force dedicated to this form of warfare and with whom I spent most my career. But it was also true of the multitude of soldiers, marines, Navy SEALs, and civilian special operations I would serve with during my time in uniform.

Observation 2: Irregular warfare missions require irregular warfare campaigns. This observation is almost a tautology, but the United States was successful in irregular war when it developed campaigns that were appropriate for the context and the type of adversary that we faced. Effective irregular warfare campaigns reflect a deep understanding of the local context and adversary, set planning horizons that are appropriate for the challenge, and rely on local solutions for local problems. These facts are widely understood, but the United States has struggled to develop military campaigns appropriate for irregular war.

Observation 3: The U.S. military is not well organized for irregular warfare campaigns. The United States has never maintained a standing capability for designing, conducting, or overseeing irregular warfare campaigns, and the only fully developed irregular warfare capability was the Office of Strategic Services, which was disestablished at the end of World War II. The prevailing assumption has been that the headquarters and formations built for traditional war could be made to work in irregular war. This reliance on the same theater- and campaign-level headquarters for conventional and irregular warfare has had clear consequences for our efficacy in irregular warfare. For one, these headquarters are not prepared to develop or execute irregular warfare campaigns or to defend against those of our adversaries. These headquarters simply lack the expertise and necessary experience in irregular warfare, as well as an appreciation of the important role of the host country in such a campaign. Resources—money, equipment, and personnel—are frequently viewed as a substitute for time or a way to overcome host-nation resistance and erode the legitimacy of our efforts and our host-nation partners. As a result, I watched these conventional formations struggle to translate tactical successes into something more enduring in nearly every population-centric campaign in which I was involved.

Observation 4: The United States lacks the concepts, doctrine, and canon necessary to be effective in population-centric conflicts. Today, we still do not have the concepts, doctrine, and canon (what constitutes good or bad thinking, practices, and policy approaches) to achieve our stated goals in population-centric conflicts. Population-centric conflicts cannot be fought with military concepts and doctrine designed for the physics of conventional war and instead require approaches that blend anthropology, economics, history, sociology, and an understanding of when and where the reality of the physics of war applies. We have retained (and continue to develop) substantial thinking on how to operate in these conflicts at the tactical level, but efforts to develop the type of thinking necessary to achieve strategic success in irregular warfare have been less successful. The problem, in my view, is that there is no proponent within the U.S. government to drive investment in the fundamentals necessary for maturing our irregular warfare capability.

Observation 5: There is insufficient professional military education for irregular warfare. The United States military does not provide its officers with the education necessary to be effective in irregular warfare. This affects our officers at the vanguard of our tactical-level irregular warfare efforts (e.g., Special Forces) but is particularly significant for other officers, including special operators with a background in more-traditional warfare (e.g., Rangers), who are often ill-equipped to lead irregular warfare–focused formations above the tactical level. There is a need to develop a service-like understanding and capability about this form of war, with a professional military education mechanism designed to provide a segment of the officer corps with expertise in the practice of irregular war up to the most-senior ranks.

Observation 6: The U.S. national security enterprise is structured to fail in population-centric conflicts. The U.S. national security enterprise within the Executive Branch is a legacy operating system designed to fight the Great Power competitions of the 20th century. A consequence is that it is inadvertently structured to fail in population-centric conflicts. For one, there is no proponent for irregular warfare in any of the five services in the Department of Defense, the Department of State, the Central Intelligence Agency, National Security Council, or anywhere else within the Executive Branch of the U.S. government.

But perhaps even more importantly, there is no superstructure to allow the United States to design and implement the whole-of-government, and potentially whole-of-society, solutions that are necessary for efficacy in this form of conflict.

Recommendations

There is no question that the United States has faced persistent difficulty in achieving strategic objectives in population-centric campaigns. It is my belief that we could have done much better in these conflicts, and will do much better in future such conflicts, if the United States reorganizes itself to take irregular warfare seriously. Yet it is unlikely that the needed reform will come from within. In the wake of Vietnam, the U.S. military deliberately turned its back on irregular warfare, vowing never to fight that kind of war again. There is a real risk that the same thing will happen in the future.

In the wake of the coordination failure that led to the failed Operation Eagle Claw and the intelligence failure that led to September 11, it took action by Congress and the support of the President to drive the reforms that we needed to maintain lethality against our modern adversary. I believe that Congress and the President will need to act again if we hope to develop the American way of irregular war that we need to provide a proactive defense and the offensive potential to destabilize our Great Power adversaries.

My three recommendations for how we can do better in these conflicts are therefore drastic and detail approaches that the U.S. Congress, President, and a group of well-financed and concerned citizens could take to force the needed reforms of the U.S. national security enterprise. My concern is that, despite our acknowledged failures, there seems to be a prevailing belief that the United States can simply retool whenever an irregular warfare capability is needed. But, in my experience, our irregular warfare capability has been insufficient even when at its peak. In my view, and based on the conversations I have had over the past four years while developing this memoir, such drastic measures as that taken in Goldwater-Nichols, Nunn-Cohen, or Collins-Lieberman are needed if we want the national security enterprise to be prepared for this form of warfare. These recommendations are interdependent, with each necessary but insufficient by itself to achieve the needed changes in America's irregular warfare capabilities.

Recommendation to Congress: Mandate a forward-looking, independent review of U.S. strategic failure in population-centric conflicts. It is past time for Congress to get involved and demand a thorough review and public accounting of U.S. performance in these population-centric conflicts. To do so, Congress should empanel a bipartisan commission with the mandate of making recommendations on how the United States can improve its policies, strategies, and campaigns in these conflicts to prepare for the ongoing irregular contests with Great Power, regional, and nonstate adversaries. This commission would assess U.S. performance in achieving strategic objectives in population-centric conflicts of all types, with the intent of diagnosing why the U.S. national security enterprise has failed to deliver strategic success in many of these conflicts and how it can be improved.

The intent of this systematic review would be to produce recommendations on the level of Goldwater-Nichols, Nunn-Cohen, or Lieberman-Collins for (potential) necessary changes in funding, structures, and authorities for irregular warfare. These previous efforts demonstrated the critical role of the U.S. Congress in forcing change in the wake of strategic failure by the United States, compelling the U.S. national security community to form U.S. Special Operations Command and the National Counterterrorism Center. However, there is a major difference between irregular warfare and these previous congressionally mandated reforms, in that we do not have a systematic understanding of why we are failing in these conflicts.

Recommendation to the President: Reorganize the Executive Branch around the security challenges of the 21st century. Maturing the American way of irregular war will very likely require reform of the Executive Branch. Although the recommended congressional study (if implemented) might provide somewhat different recommendations, I believe that the President of the United States and Executive Branch staff ought to examine three options for how the United States could better organize for success in these conflicts:

1. Create a contemporary version of the World War II–era Office of Strategic Services by establishing a cabinet-level organization with primary responsibility for paramilitary, influence, and special warfare operations.
2. Create a separate service within the Department of Defense that is missioned to own irregular warfare or a new service under the Department of the Army that is focused on irregular warfare, building off the model that is currently employed by the Navy and Marine Corps.

3.   Divide U.S. Special Operations Command into two four-star functional combatant commands, with one focused on the national priority missions of counterterrorism and countering weapons of mass destruction (in close alignment with the command's current focus) and the second focused on irregular warfare.

In each of these options, the new superstructure for irregular warfare would incorporate military, foreign service, and intelligence professionals in prominent (if not leadership) roles. Each of these options also recognizes that research and analysis will be necessary to develop concepts for America's use of irregular warfare. Maintaining a persistent but low-profile global network of irregular warfare professionals will be critical to providing situational awareness and access, influence, and indigenous options and enhancing America's use of influence operations.

Recommendation to concerned citizens: Establish an institution outside government dedicated to understanding American irregular warfare. Arguably, our victory in the Cold War resulted as much from America's ability to harness its intellectual capacity as it did from its industrial might. The United States has not had—again, in my view and based on my own experience—the same quality of expertise for irregular warfare. I believe that an independently funded center, or a public-private center supported by both Congress and private citizens, at a university or think tank dedicated to the study of American irregular warfare would provide our country three necessary capabilities. The first is a continuous and independent critique of U.S. capabilities, policies, and strategies in irregular warfare to determine how the United States is performing in its many irregular engagements and how it might do better in this type of war. Second, the institution would provide a stable of professionals who are experts in the contemporary use of irregular warfare, both how our adversaries apply this form of warfare and how the United States can deploy irregular warfare defensively and offensively to contest these adversaries, whether state or nonstate. The third would be to capture and analyze irregular warfare experiences, providing a publicly available record of our successes and failures and thus serving as a bridge between irregular warfare practitioners and the American people.

Conclusion

Maturing the American way of irregular war is critical for the United States to prevail in conflict and competition in the 21st century. We need to be as agile as our adversaries, and we must have the capability to be both reactive and proactive, allowing us to simultaneously counter our adversaries' irregular threats and go on the offensive. This offensive capability would give the United States an added ability to deter our enemies by expanding the competitive space of the U.S. military. The Achilles' heel of our authoritarian adversaries is their inherent fear of their own people; the United States must be ready to capitalize on this fear.

An American way of irregular war will reflect who we are as a people, our diversity, our moral code, and our undying belief in freedom and liberty. It must be both defensive and offensive. Developing it will take time, require support from the American people through their Congress, and is guaranteed to disrupt the status quo and draw criticism. It will take leadership, dedication, and courage. It is my hope that this study encourages, informs, and animates those with responsibility to protect the nation to act. Our adversaries have moved to dominate in the space below the threshold of war. It will be a strategy built around an American way of irregular war that defeats them.

Acknowledgments

I would first like to thank Daniel Egel, my coauthor, and now good friend, for making this work a reality. His unflagging editorial support and diligent research turned a collection of old war stories into a set of recommendations that, we hope, will begin the work of correcting a shortfall in America's defense. The project was funded by the Smith Richardson Foundation and undertaken with the support and encouragement of Nadia Schadlow and Linda Robinson, to whom I am grateful for giving me this opportunity. I have known Linda for many years and continue to be inspired by her extensive research into irregular warfare and special operations.

Further, we want to give special thanks to Ambassador (retired) Ryan Crocker, Andrew Liepman, and Colonel (retired) David Maxwell for providing advice, guidance, and insight throughout the project. Lieutenant General (retired) Ken Tovo provided excellent feedback on an early draft, and both Mike Vickers and Michael Sheehan provided superb guidance and direction at an early stage of the project. We are also appreciative of input from Captain Edward Dawson and Lance Golat, who provided very useful feedback at a critical juncture in the project. The study benefited enormously from detailed peer reviews by Ben Connable of RAND and James Kiras at the Air University's School of Advanced Air and Space Studies.

Many of the key insights in this study are based on a workshop held with a group of visionaries from the U.S. national security world. We would like to extend our special thanks to each of these participants—Adam Barker, Ambassador (retired) Beth Jones, Seth Jones, Steve Metz, Phillip Mudd, Lieutenant General (retired)

John Mulholland, Paul Nevin, Mike Noblet, Jeanne Pryor, Ambassador (retired) Marcie Ries, Michael Sheehan, Richard Tamplin, Paul Tompkins, Mike Vickers, Theresa Whelan, Jonathan White, and Ken Yamashita—who provided both oral and written feedback that guided the thinking and design of the project. In addition, I would like to thank the interviewees who agreed to share their views with us, which included a variety of national security–focused analysts and representatives from the Department of Defense, U.S. Agency for International Development, and U.S. intelligence community.

Abbreviations

CENTCOM	U.S. Central Command
CIA	Central Intelligence Agency
CJSOTF	Combined Joint Special Operations Task Force
CJSOTF-A	Combined Joint Special Operations Task Force–Afghanistan
DEA	Drug Enforcement Administration
FARC	Fuerzas Armadas Revolucionarias de Colombia
NATO	North Atlantic Treaty Organization
SOCCENT	Special Operations Command Central
SOC Forward	Special Operations Command Forward
SOCOM	U.S. Special Operations Command
SOCSOUTH	Special Operations Command South
SOUTHCOM	U.S. Southern Command
TRADOC	U.S. Army Training and Doctrine Command
UMOPAR	Unidad Movil Policial para Areas Rurales
USAID	U.S. Agency for International Development
USASOC	U.S. Army Special Operations Command

CHAPTER ONE

Introduction

My professional military life began in the shadow of America's defeat in Vietnam. The fall of Saigon, which happened at the end of my first year at the U.S. Military Academy at West Point, was shocking to our country's leaders, all members of the greatest generation. The men who stormed the beaches at Normandy against the Nazis, captained patrol boats in the Pacific to defeat the Japanese, and learned under Dwight Eisenhower, Omar Bradley, and Douglas MacArthur had failed in a war against a third-world communist proxy. And America had failed in its efforts to stop the spread of communism in Southeast Asia.

The imagery of failure captured by the helicopter rooftop ballet, carrying away as many of the desperate as possible,[1] and North Vietnamese Army troops parading through the capital of our defeated ally were the mental frames under which I started my apprenticeship in war. At the time, I was concerned with the very naive notion that an army that had not known defeat had suffered its first lost war. On hearing the news, my English professor, a well-respected infantry cap-

---

[1] On April 29, 1975, U.S. Ambassador Graham A. Martin ordered the evacuation of U.S. forces from Vietnam. With the airport damaged—by a combination of North Vietnamese artillery fire and aerial bombing, the latter which was led by a renegade South Vietnamese pilot—the evacuation was done by helicopter, with more than 7,000 Americans and South Vietnamese evacuated in under 24 hours. This evacuation became the "ultimate symbol of the failure of US policy in Southeast Asia," with the media presenting "hundreds of wrenching scenes—tiny boats overcrowded with soldiers and family members, people trying to force their way onto the US Embassy grounds, Vietnamese babies being passed over barbed wire to waiting hands and an unknown future" (Walter J. Boyne, "The Fall of Saigon," *Air Force Magazine*, April 2000).

tain, decorated for valor, broke down in class. But it was not of course for the stain on the Army's record but for what he judged was a waste of life and treasure and the needless loss of friends and subordinates.

The adversary we faced in Vietnam, an adversary who operated and sheltered among the indigenous people, was to be the most pernicious and persistent security threat to the United States throughout my career. It was not the nuclear threat from the Soviets nor the misguided conventional use of arms by dictators and tyrants that would threaten the American way of life. The threat would be instead from adversaries who relied on nonconventional means, leveraging population-centric approaches, such as resistance, rebellion, insurgency, terror, and civil war. It was these population-centric campaigns that would prove a steady drain on American wealth, influence, and confidence.

The United States has struggled in these population-centric contests. During the nearly four decades of my career, U.S. combat experiences revealed stunning military capabilities and repeated tactical successes, yet an inability to successfully conclude contests in which our enemy could operate and shelter among indigenous people. The troubled campaigns in Afghanistan and Iraq, against the Islamic State, and against irregular forces in Somalia, Yemen, and Libya have shown that the United States is unable to translate tactical successes into something more enduring in population-centric campaigns.[2] Neither America's current irregular warfare capability, which has extremely limited ability above the tactical level, nor its more mature conventional forces have proven effective.

This analytical memoir is an effort to describe how improvements in America's irregular warfare capability might make the United States more effective in population-centric conflicts. The analysis draws extensively on my career in the U.S. Army as a student and practitioner of irregular warfare, the bulk of which was spent with the U.S. Army Special Forces. Time and time again I saw tremendous success at the

---

[2]   General Rupert Smith made a similar argument, concluding that the "old paradigm was that of interstate industrial war," the "new one is the paradigm of war amongst the people," and "conventionally formulated military forces" do not succeed in these new wars (Rupert Smith, *The Utility of Force: The Art of War in the Modern World*, New York: Knopf, 2007, pp. 5–6).

tactical level in these population-centric contests, in Europe, in Latin America, and in the Middle East. But, in each case, we lacked the operational and institutional structures to embed our tactical proficiency into irregular warfare campaigns that reflected a sufficient understanding of our adversary and the context of these fights. The result was that we almost always failed to achieve strategic objectives, despite never failing in a single mission.

My conclusion, based on my 37-year career in the U.S. Army and the four-year research process that led to this manuscript, is that the United States must develop an American way of irregular war. This capability should complement the peerless traditional warfare–focused American way of war, providing us a counter for adversaries who employ strategies below the threshold of traditional war. I do not hope to be able to understand the full range of capabilities that we must develop, which is beyond the scope of any one person, but rather to diagnose the challenge and provide a pathway that the United States could take to develop these capabilities.

An American Way of Irregular War

The American approach to warfare, which focuses on the annihilation of enemy military forces through the "awesome destructive power that only a fully mobilized and highly industrialized democracy can bring to bear,"[3] has proven remarkably effective in conventional war. This is demonstrated dramatically by historical American successes in the two World Wars, the First Gulf War, and in the beginning of our campaign in Iraq (among others). Yet this approach, which Russell Weigley famously coined the *American way of war*,[4] has proven much less effective in population centric contests. Its focus on the annihilation of an enemy's military forces, however effective for conventional warfare, has

---

[3]   Max Boot, "The New American Way of War," *Foreign Affairs*, 2003.

[4]   Russell F. Weigley, *History of the United States Army*, New York: Macmillan, 1967.

proved insufficient in conflicts in which the "aftermath" of fighting is as critical to strategic success as is the fighting.[5]

The American way of irregular war, in contrast, achieves strategic effects against adversaries through attrition rather than annihilation. It is an approach to warfare that emphasizes the importance of local partnerships and gaining legitimacy and influence among local populations.[6] It typically involves extended-duration campaigns that blend U.S. military and civilian capabilities, often with a prominent role for military or civilian special operations capabilities,[7] combining overt and covert capabilities.[8] Strategic success is determined based on

---

[5] Antulio J. Echevarria II, "What Is Wrong with the American Way of War?" *Prism*, Vol. 3, No. 4, 2012, p. 113.

[6] This definition is based on my own experience and the consensus of esteemed participants in a workshop organized for this project, the vast majority of whom emphasized either the importance of indigenous forces or the goal of influencing a population. The (late) Michael Sheehan summarized the perspectives of the other participants very elegantly, when he concluded that irregular warfare was the use of "asymmetric methods to . . . undermine [an adversary's] political legitimacy and ultimately gain political victory" (Michael Sheehan, private communication). This definition encapsulates both the congressional and Department of Defense definitions. The 2018 National Defense Authorization Act defined *irregular warfare* as the activities conducted "in support of predetermined United States policy and military objectives conducted by, with, and through regular forces, irregular forces, groups, and individuals participating in competition between state and non-state actors short of traditional armed conflict" (Pub. L. 115-91, National Defense Authorization Act for Fiscal Year 2018, December 12, 2017). The Department of Defense defines *irregular warfare* as "[a] violent struggle among state and non-state actors for legitimacy and influence over the relevant population(s)" (U.S. Department of Defense, *Irregular Warfare [IW] Joint Operating Concept (JOC)*, Version 1.0, Washington, D.C., September 2007, p. 1).

[7] James Kiras describes how—in long-duration campaigns—an effective blend of U.S. special operations, military, and civilian capabilities can defeat adversaries through attrition (James Kiras, *Special Operations and Strategy: From World War II to the War on Terrorism*, New York: Routledge, 2006; James Kiras, "A Theory of Special Operations: 'These Ideas Are Dangerous,'" *Special Operations Journal*, Vol. 1, No. 2, 2015).

[8] Andrew Liepman, who served at the highest levels of the Central Intelligence Agency (CIA), was one of the few workshop participants who indicated that he could not provide a concise definition, which he indicated "contributes to the problem" that we face in this form of warfare. From his perspective: "It's counterinsurgency, it's information operations, it's covert action—it isn't tanks, aircraft, artillery, or ships—it's not force-on-force combat, the kind of thing we planned for against the Soviets, what we did in Iraq in 1991. But it depends on those things to succeed. Irregular warfare is through partners and for partners. Virtually

the ability to control and influence populations rather than occupy territory, and success requires approaches that blend anthropology, economics, history, and sociology rather than only the "physics" necessary for conventional warfare.[9] The American way of irregular war is also a form of warfare in which the United States, similar to many other states that have dominated in conventional warfare, has struggled because the "cultural, political, and military qualities" necessary for irregular warfare are not "traditional strengths of Americans."[10]

Although other nations, friendly and adversarial, have developed or are developing analogous capabilities, this approach to warfare is uniquely American. In the same way that the conventionally focused American way of war is defined by America's technical and industrial capacity and technological edge, the American way of irregular war is tied to our notions of religious pluralism, democracy, and, above all, human rights. And although the American way of war protects us against near-peer powers and guarantees the lanes of global commerce, the American way of irregular war protects our way of life by both promoting our worldview and giving people the tools to realize the same opportunities that we have had.[11] In this way, it also differs

---

everything we're doing in Iraq and Afghanistan is irregular—using force mostly indirectly in support of another entity (the Afghan army, Syrian opposition, the Iraqi government) but in a fight, that requires more than military victory to succeed" (Andrew Liepman, personal communication).

[9]   Paul Tompkins aptly concluded that traditional warfare is fought in the physical domains and is all about occupying and controlling the enemy's space to force your will on the enemy, whereas irregular warfare is fought in the minds of the people (Paul Tompkins, personal communication). Rob Pyott had a similar description, describing irregular warfare as "warfare by means other than a raid; . . . generally, it is a military action that has an impact on a key population (enemy or third party)" (Rob Pyott, personal communication).

[10] Colin S. Gray, "Irregular Warfare: One Nature, Many Characters," *Strategic Studies Quarterly*, Vol. 1, No. 2, 2007.

[11] David Maxwell captured this eloquently, writing that irregular warfare "consists of competition over two ideas: (1) the national interest to maintain a stable international nation-state system based on respect for and protection of sovereignty and (2) the fundamental human right of people to seek self-determination of government" (David Maxwell, personal communication).

significantly from the irregular warfare approaches employed during America's formative years.[12]

For the United States, the long-term goal of irregular warfare efforts is frequently to help a host nation address the root causes of instability (e.g., economic, judicial, social, religious), and the rest of our efforts—neutralizing the enemy, defending the population, and mobilizing the populace—are just supporting tasks to buy time.[13] Thus, although our competitors focus on using their irregular warfare capabilities to sow chaos and instability, the United States often uses its own to build and stabilize. However, when necessary, the United States is prepared to do what it is necessary to "free the oppressed," hence the *de oppresso liber* motto of my Special Forces regiment. Anything less would be anathema to our core values as Americans.

The intent of this study is to describe how the United States might mature the American way of irregular war, which I believe is critical if the next generation hopes to avoid the types of costly failures that we have seen in Vietnam, Afghanistan, Iraq, and many other population-centric conflicts. My goal is to describe the types of reform necessary for the United States to develop the concepts, supporting doctrine, organizations, and training necessary to increase the incidence and durability of success in these conflicts. These reforms will give us the world-class irregular warfare capability, mirroring our conventional supremacy, that we need.

---

[12] John Grenier, *The First Way of War: American War Making on the Frontier, 1607–1814*, Cambridge: Cambridge University Press, 2008.

[13] This characterization is based on a very apt observation from Lieutenant General (retired) Ken Tovo about our approach in Iraq. He divided our counterinsurgency efforts into four elements: (1) Neutralize the insurgents, an offensive task typically done with a host-nation partner force that was being trained to be competent, professional, and ethical; (2) secure the population, a defensive task; (3) mobilize the population, which typically involved a blend of methods to encourage the population to support the host-nation government; and (4) balance development to address the causes of instability. The critical observation was that the only nonmilitary activity was the main effort and that elements 1–3 are only supporting efforts to buy time for this main effort.

Prevalence of Population-Centric Conflict

The population-centric conflicts that are the focus of America' irregular warfare capability—insurgency, terrorism, resistance, subversion, and civil war—are the most prevalent form of conflict facing the United States. Their growing prevalence is illustrated in Figure 1.1—which reports the number of governments directly involved in population-centric warfare, as compared with traditional and colonial wars, for each year since World War II. In 2016 alone—in addition to the major multinational irregular wars in Afghanistan, Iraq, Mali, Somalia, and Yemen—there were 47 population-centric conflicts involving 49 governments, including the contest between Ethiopia and the Oromo Liberation Front, Turkey's struggles with Kurdish separatists, and 14 separate contests involving the Islamic State.[14]

The human cost of these contests has been tremendous. More than 6.5 million souls have died, representing more than 70 percent of all conflict-related deaths during that time, as illustrated in the right panel of Figure 1.1.[15] Two major conflicts accounted for many of these deaths, as an estimated 1.2 million died in the Chinese Civil War and an additional 1.6 million died in the Vietnam War. But nine other significant conflicts—those with more than 100,000 deaths, including the Soviet-Afghan War, Khmer Rouge, Greek Civil War, Lebanese Civil War, and Ethiopian Civil War—claimed nearly 2 million lives, and the more than 200 other population-centric conflicts together claimed 2 million more.

---

[14] Based on Nils Petter Gleditsch, Peter Wallensteen, Mikael Eriksson, Margareta Sollenberg, and Håvard Strand, "Armed Conflict 1946–2001: A New Dataset," *Journal of Peace Research*, Vol. 39, No. 5, 2002; and Kristine Eck and Thérése Pettersson, "Organized Violence, 1989–2017 and the Data Generation Process," *Journal of Peace Research*, Vol. 55, No. 4, 2018. See the note to Figure 1.1 for a further discussion of data and definitions.

[15] See the note to Figure 1.1 for a discussion of the data underlying this estimate.

Figure 1.1
Prevalence of Population-Centric Warfare Since World War II

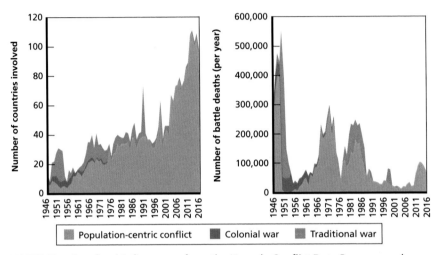

NOTES: The data for this figure are from the Uppsala Conflict Data Program and Centre for the Study of Civil Wars, International Peace Research Institute, Oslo, which record every conflict with "at least 25 battle-related deaths in a calendar year" (Lotta Themnér, *UCDP/PRIO Armed Conflict Dataset Codebook*, Version 18.1, Uppsala, Sweden, and Oslo, Norway: Uppsala Conflict Data Program and Centre for the Study of Civil Wars, International Peace Research Institute, 2018, p. 1). *Population-centric conflict* includes all internal armed conflicts between the government of a state and an internal opposition group, with or without external support; *colonial wars* are conflicts in which a government is fighting to retain external territory; and *traditional wars* are conflicts between two or more states (Themnér, 2018, pp. 9–10). The left panel reports our estimates of total number of governments involved in at least one of each type of conflict for a given year, based on the data set in Gleditsch et al., 2002; Kristine Eck and Therése Pettersson, "Organized Violence, 1989–2017 and the Data Generation Process," *Journal of Peace Research*, Vol. 55, No. 4, 2018. The right panel reports estimates of the total number of battle deaths—military and civilian—based on the combination of International Peace Research Institute's Battle Deaths Dataset (Bethany Lacina and Nils Petter Gleditsch, "Monitoring Trends in Global Combat: A New Dataset of Battle Deaths," *European Journal of Population*, Vol. 21, Nos. 2–3, 2005), which provides data from 1946 to 2008, and the Uppsala Conflict Data Program's UCDP Battle-Related Deaths Dataset (Pettersson and Eck, 2018), which provides data from 1989 to 2017. The numbers reported include an amalgamation of the (1) "best estimate" reported in the Uppsala Conflict Data Program data, (2) best estimate in the International Peace Research Institute's data, and (3) lowest available estimate reported in the International Peace Research Institute's data—these estimates are reported in our preferred order, with the lower-ranking estimate used only when the previous are not available.

America's Irregular Warfare Misadventures

The United States has been involved in a great many of these post–World War II population-centric contests. As illustrated in Figure 1.2, these conflicts have included operations in nearly 50 countries, from China (Chinese Civil War) to East Germany (Cold War) to El Salvador (Salvadoran Civil War). There were also missions that lasted decades (drug war in Bolivia, 1986–2008) and others that lasted only a few months (Lebanon crisis of 1958). And these conflicts involved a great many types of U.S. forces, including U.S. civilian agencies (e.g., Drug Enforcement Administration [DEA], Department of State, CIA) and conventional and special operations elements from the Department of Defense.

U.S. involvement in these conflicts has relied on America's irregular warfare capability. In some cases, this irregular warfare capability has emphasized U.S. unilateral action, as was the case during the Lebanon crisis of 1958, when U.S. marines and soldiers occupied Beirut following a coup d'état; Operation Just Cause in 1989, which removed Panamanian President Manuel Noriega from office to safeguard U.S. lives and interests in the Canal Zone; and the Osama bin Laden raid in Pakistan in 2011. However, the vast majority of U.S. involvement in these conflicts has relied on partnerships and relationships with locals, including support to partisans seeking political change in line with U.S. interests (Tibet Resistance, Overthrow of Taliban) and covert support to surrogates (Cuba). In recent years, support to partner governments that face domestic insurgency, terrorism, resistance, and subversion has featured prominently, with the United States building domestic capabilities to resist these challenges; this approach is exemplified by ongoing U.S. support to the Afghan and Iraqi governments.

Yet, regardless of the approach, strategic victories in these irregular wars have proved elusive, particularly for a country that still gauges military success by the unconditional surrender after World War II. We celebrate the tactical brilliance of the Osama bin Laden raid, the overthrow of the Taliban by horseback Special Forces and exotic local forces, and the heavy metal Thunder Run through Baghdad. But the troubled campaigns in Afghanistan in Iraq and against the Islamic

Figure 1.2
U.S. Involvement in Population-Centric Conflicts Post–World War II

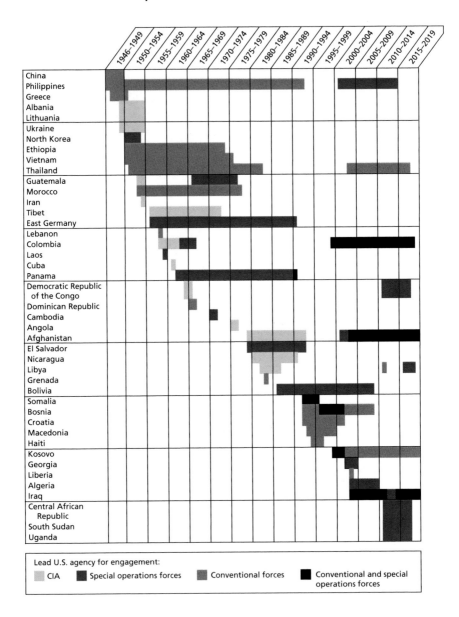

Figure 1.2—Continued

SOURCES: This figure is derived from a multitude of sources, including Jennifer Kavanagh, Bryan Frederick, Alexandra Stark, Nathan Chandler, Meagan L. Smith, Matthew Povlock, Lynn E. Davis, and Edward Geist, *Characteristics of Successful U.S. Military Interventions*, Santa Monica, Calif.: RAND Corporation, RR-3062-A, 2019; Gleditsch et al., 2002; Eck and Pettersson, 2018; Russel Crandall, *America's Dirty Wars: Irregular Warfare from 1776 to the War on Terror*, New York: Cambridge University Press, 2014; D. Jones, *Ending the Debate: Unconventional Warfare, Foreign Internal Defense, and Why Words Matter*, Fort Leavenworth, Kan.: U.S. Army Command and General Staff College, 2006; William J. Daugherty, *Executive Secrets: Covert Action and the Presidency*, Lexington: University Press of Kentucky, 2004; Peter Harclerode, *Fighting Dirty: The Inside Story of Covert Operations from Ho Chi Minh to Osama Bin Laden*, London: Cassell and Company, 2001; James Stejskal, *Special Forces Berlin: Clandestine Cold War Operations of the US Army's Elite, 1956–1990*, Philadelphia, Pa.: Casemate Publishers, 2017; Dennis M. Rempe, *The Past as Prologue? A History of U.S. Counterinsurgency Policy in Colombia*, 1958–66, Carlisle, Pa.: Strategic Studies Institute, U.S. Army War College, 2002; Piero Gleijese, *Conflicting Missions: Havana, Washington, and Africa, 1959–1976*, Chapel Hill, University of North Carolina Press, 2002; Lorenzo Vidino, "How Chechnya Became a Breeding Ground for Terror," *Middle East Quarterly*, Vol. 12, No. 3, Summer 2005; Armando J. Ramirez, *From Bosnia to Baghdad: The Evolution of US Army Special Forces from 1995–2004*, Monterey, Calif.: Naval Postgraduate School, 2004; Linda Robinson, *Masters of Chaos: The Secret History of the Special Forces*, New York: PublicAffairs, 2001; Hy Rothstein, "Less Is More: The Problematic Future of Irregular Warfare in an Era of Collapsing States," *Third World Quarterly*, Vol. 28, No. 2, 2007; Thomas K. Adams, *US Special Operations Forces in Action: The Challenge of Unconventional Warfare*, Portland, Ore.: Frank Cass Publishers, 1998; Gordon L. Rottman, *Mobile Strike Forces in Vietnam, 1966–70*, London: Osprey, 2013; Mark Moyar, Hector Pagan, and Wil R. Griego, *Persistent Engagement in Colombia*, Tampa, Fla.: Joint Special Operations University, July 2014; and John Barry, "America's Secret Libya War: U.S. Spent $1 Billion on Covert Ops Helping NATO," *Daily Beast*, July 13, 2017.
NOTES: This figure provides a representative, but not comprehensive, list of U.S. involvement in irregular wars. It indicates the approximate timing of U.S. engagement and the U.S. military (or paramilitary force, in the case of the CIA) that led the military component of the engagement. Most of these engagements also included a prominent role by other U.S. civilian agencies.

State and al Qaeda; problematic outcomes in Libya, Somalia, Yemen, and Pakistan; and the seemingly impotent responses to Russian, Chinese, and Iranian foreign adventures demonstrate the inability of the United States to achieve strategic outcomes in these contests.

During the final decade of my career, while commanding special operations units at the operational and institutional levels of America's irregular warfare capability, I began to believe that our inability to achieve strategic effects was partly because the United States was not properly organized to conduct irregular warfare above the tactical level. We were almost always successful at the outset of contests in which we chose direct military confrontation and the application of military firepower was dominant, but we failed to consolidate the victory in the human-centric portion.[16] Our understanding of the enemy and local populations was flawed, and we lacked the organizational structures to properly design and execute the types of long-duration campaigns needed. The result was military campaigns that failed to deliver at the strategic level and bad policy recommendations to senior leaders.

All evidence suggests that our adversaries recognize this vulnerability and have learned more from our failures than we have. They recognize the superiority of America's technology and skills at conventional maneuver and instead fight America and its allies through nonconventional means, including resistance, rebellion, insurgency, subversion, and terror. Islamic revolutionaries use terror, such as the attacks on September 11 or the bombings in Madrid, to either scare us off or draw us into war. Russia uses an artful mix of powerful conventional formations along with irregulars, disinformation, and the Russian diaspora. Iran aggressively uses its Lebanese Hezbollah partner, intelligence service, the Quds Force of Iran's Islamic Revolutionary Guard, and Shia militias to meddle abroad. And although the Chinese government primarily uses its economic clout to influence affairs

---

[16] For some of these conflicts, we decided thankfully not to get involved with conventional forces, such as Yemen and Somalia, but they still proved problematic. Yemen is perhaps a particularly good example, as we underestimated the Houthi and Iranian threat, perhaps purposely so to protect the Joint Comprehensive Plan of Action, but the outcome was not what we set out to do. That is also a reality of this form of war—we won't win them all. The goal is to lose less badly, be less surprised, and be wiser about the use of force.

in other countries, it has increasingly turned to joint military training events, special forces–type advisory missions, and overseas basing to pressure others.

My experiences, world events since the Vietnam War, and the writings of our Great Power adversaries lead me to believe that irregular warfare will remain the most prevalent form of conflict facing the United States for many years to come. Both state and nonstate adversaries, leveraging approaches that can best be described as irregular warfare, have already imposed costs on the United States. Tens of thousands of American citizens and servicemembers have been killed or maimed, and trillions of dollars have been spent. Perhaps most concerning strategically, this approach has negatively affected the readiness of American military forces, leaving the United States potentially ill-equipped to respond to real conventional threats, including against possible existential threats of the future.[17]

Analytical Approach

The format for this study follows that pioneered by Lieutenant General (retired) Glenn Kent in *Thinking About America's Defense: An Analytical Memoir*.[18] I, like Kent, intend for this study to be neither a memoir nor a biography of my career, as it seems that neither of us is "really interested in recounting the events of [our lives], fascinating though they are."[19] The study is instead intended to be a review of my own process of discovery, tracing the significant moments in my career when I saw the strengths and weaknesses of the emergent American way of irregular war. My intent is to describe what I learned along the way,

---

[17] Even analysts who argue that U.S. military readiness has *not* been significantly affected do agree that America's armed forces "remain better prepared for counterinsurgency and stabilization missions than for high-end warfare" (Michael O'Hanlon, *The State of U.S. Military Readiness*, Washington, D.C.: Brookings Institution, August 15, 2016).

[18] Glenn A. Kent, with David Ochmanek, Michael Spirtas, and Bruce R. Pirnie, *Thinking About America's Defense: An Analytical Memoir*, Santa Monica, Calif.: RAND Corporation, OP-223-AF, 2008.

[19] Kent et al., 2008, p. iii.

whether through my own experiential learning—or that of colleagues and friends—or through the multitude of excellent research and analysis, some of which I had the opportunity to encourage and even support, that has been produced during these past 40 years.

This analytical memoir approach has two primary characteristics. The first, which has been illustrated throughout this introductory chapter, is that it draws heavily on my experiences over a career at the tactical, operational, and institutional levels of America's irregular warfare capability. This autobiographical approach is widely recognized as an "extraordinarily valuable vehicle,"[20] and it is perhaps particularly appropriate for the subject matter given the paucity of detailed and systematic accounts of how the United States has fought its irregular wars.[21]

The second characteristic, which is the key way in which this analytical memoir differs from a traditional autobiography, is that I use a blend of primary and secondary sources to test, validate, and refine the insights and observations. This approach is widely recognized as a best practice among academics who use personal experience as a qualitative research tool.[22]

The primary data collection included consultations with 24 researchers and practitioners (both retired and active) from the Department of Defense, Department of State, U.S. Agency for International Development (USAID), and CIA. In addition, I hosted a senior-level

---

[20] Mark P. Freeman, "Autobiography," in Lisa M. Given, ed., *The SAGE Encyclopedia of Qualitative Research Methods*, Vol. 2, Thousand Oaks, Calif.: SAGE Publications, 2008. Freeman cautions that autobiographies can be "capricious, error filled, and distortive," which is something that I take seriously and is the reason I try to use all available information to validate my own experiences.

[21] For a relevant discussion of the strengths and limitations of autobiography for this type of analysis, see Morgan Brigg and Roland Bleiker, "Autoethnographic International Relations: Exploring the Self as a Source of Knowledge," *Review of International Studies*, Vol. 36, No. 3, 2010.

[22] This approach has been identified as a good practice in using either autobiography (e.g., see the "Qualitative Research as Autobiography" section in Steven J. Taylor, Robert Bogdan, and Marjorie DeVault, *Introduction to Qualitative Research Methods: A Guidebook and Resource*, 4th ed., Hoboken, N.J.: Wiley, 2015, pp. 140–142) or autoethnography (Brigg and Bleiker, 2010) as research tools.

workshop with 17 experts—former uniformed practitioners, intelligence officers, career diplomats, development professionals, and policy experts.[23] In this study, this primary data collection is used mainly to inform the observations and the recommendations that come at the conclusion of the memoir.

In terms of secondary sources, it was simply not possible to provide a comprehensive review of the extremely broad and diverse literature on irregular warfare and America's approach to irregular warfare. I instead took a two-pronged approach for engaging with the literature. First, I try to include what I believe to be seminal texts for understanding American irregular warfare, which are the texts that have guided my thinking and the thinking of the Department of Defense throughout my career. Second, for each episode of my career that I examine in the text, I try to provide as comprehensive a review of the literature as possible. This is easier for some of these episodes than others, as the analysis of the U.S. irregular warfare efforts in Bolivia (Chapter Three) references all academic and government reporting on the counterdrug campaign, while the literature relevant to my experience in Iraq (Chapter Eight) is too expansive to hope to include in its entirety. For this latter group, I include what the project team believes to be a representative sample of existing accounts.

Organization of the Analytical Memoir

The remainder of this memoir consists of 13 chapters. Chapters Two through Eleven are organized around the major episodes of my nearly four-decade career, with each chapter describing what I learned about the strengths and limitations of our existing concepts, doctrine, organizational structures, and training for irregular warfare. Chapter Twelve summarizes my key observations about irregular warfare at the conclu-

---

[23] This workshop asked participants to (1) diagnose why the United States has been unable to achieve durable and favorable strategic outcomes in recent campaigns, (2) evaluate whether existing security sector structures and conceptual frameworks are sufficient for understanding and countering today's nonconventional threats and those over the horizon, and (3) propose approaches for addressing the identified problems.

sion of this extended personal period of discovery, Chapter Thirteen provides recommendations for reforms that the United States should consider if we want to develop the mature American way of irregular war that I believe that we need, and Chapter Fourteen concludes with a look to the future, describing the urgency of formalizing and maturing this American way of irregular war if we hope to counter the threats we face today, as well as those just over the horizon.

The ten chapters describing the major episodes in my career can be organized into the three major stages through which my career progressed. The first stage was my experiences during the Cold War, in which irregular warfare became a central element in efforts to contain the spread of communism. This included five and a half years oriented toward general war in Europe (Chapter Two), not including military schooling—three years with 10th Special Forces Group (Airborne) planning and preparing for unconventional warfare and two and a half years commanding VII Corps' Counterintelligence and Interrogation Company, which meant frequent trips to our border posts along the Czechoslovakian and East German borders. The next three and a half years were spent focused on Latin America, in Bolivia (Chapter Three), El Salvador (Chapter Four), and Panama (Chapter Five) in a series of irregular warfare conflicts that were part counternarcotics and part the containment of communism.

The second stage was the decade following the successful conclusion of the Cold War, which I refer to as the "decade of delusion," as this was a period in which the United States began to shed much of its irregular warfare capability because of a perception, resulting from America's overwhelming success in the First Gulf War, that our conventional dominance rendered it obsolete. I spent much of this decade in the Pentagon (Chapter Six), involved in what I would later think of as my "Pentagon wars," as we fought to retain the irregular warfare capability that many of us in Special Forces knew that the nation would need again. And I concluded the decade again with 10th Special Forces, this time overseeing an expanding role for irregular warfare in Europe—specifically, in Bosnia, Georgia, and Kosovo (Chapter Seven). While others were seeing the end to conventional fights in the new world order, it was clear to me that population-centric warfare was not going way.

The final stage began with the terrorist attacks on September 11, 2001, in which the American irregular warfare capability first became a true centerpiece in our national security strategies. This stage of my career began with a prominent role in redeveloping America's unconventional warfare capability in the post–September 11 global war on terror, which culminated with a leadership role in the initial invasion of Iraq (Chapter Eight). I then had the opportunity to coordinate America's irregular warfare efforts, first in Latin America as the commander of Special Operations Command South (SOCSOUTH) (Chapter Nine) and then in the Middle East, Central Asia, and Southwest Asia as the commander of Special Operations Command Central (SOCCENT) (Chapter Ten). I concluded my career (Chapter Eleven) as the commanding general of U.S. Army Special Operations Command (USASOC), now responsible for all U.S. Army special operations.

During this last stage of my career, and particularly during my time at USASOC, I was determined to undertake a fundamental relook at how the United States fought these population-centric contests and create models and concepts on which to build better doctrine and the right forces to implement it. We made some progress, but we failed to achieve the change in thinking and behavior that I know is necessary if the United States hopes to protect itself against these threats.

I hope that this memoir may continue this progress, providing analysis that will serve as an impetus to develop and mature the American way of irregular war. The United States turned its back on irregular warfare in the wake of Vietnam and threatens to do so again today. We cannot hope to remain competitive, and to maintain our place and influence in the world, if we do not develop a capability for irregular war that, like our conventional and nuclear capabilities, can dominate on the global stage.

I recall my West Point English professor and his anguish at the fall of Vietnam as I gauge what has been achieved in the past two decades of war. I have seen the cost only too clearly, in the faces of those maimed and in the lamentations of parents, spouses, and children of those killed. I ask myself each time: Was it worth it? If we knew then what we know now, what would we have done differently? What would we have wanted in our arsenal that we did not have? Most important, were decisions grounded in a studied understanding of the type of war that policymakers intended to make?

As other Great Powers build out their conventional capabilities, they will also continue to exploit America's increasingly apparent weakness in irregular war. As the threat broadens to include now-traditional war, America needs forces that are capable and leaders who are wiser across the spectrum of conflict. To do so, America must study irregular war and develop concepts and structures to win such contests. I sincerely believe that much of the death and maiming of young Americans that I saw during my career was not inevitable and that maturing an American way of irregular war is necessary to make worthy the inevitable future sacrifice of America's sons and daughters and to regain dominance over our adversaries.

CHAPTER TWO

Irregular Warfare and the Cold War in Europe

In June of 1979, the naive, overconfident, but worldly son of an Army noncommissioned officer reported to the 10th Special Forces Group headquarters at Fort Devens, Massachusetts. Our mission at 10th Special Forces was to prepare for unconventional warfare throughout Europe if the Soviet Union were ever to attack, slipping behind enemy lines to organize, train, and fight alongside resistance fighters. I would spend three years with 10th Special Forces, and an additional three years supporting the broader Cold War mission in Europe as an intelligence officer assigned to U.S. Army Europe's VII Corps in Stuttgart.

By 1946, the containment of communism had emerged as a national policy, and irregular warfare was to be a core element of U.S. efforts. The game was on almost immediately, as we struggled against the Soviet Union to manipulate political outcomes in China (unsuccessfully) and Greece (successfully). Similar contests would emerge in Asia (e.g., Korea, Tibet, Vietnam), Africa (e.g., Algeria, Kenya), and Latin America (e.g., Bolivia, Cuba) in the following years.

These efforts were deemed necessary to avoid "the big one." Conventional war on the plains of Europe, in front of and behind the Iron Curtain, and in the era of thermonuclear weapons was an unthinkable event. These contests of the "new peace" meant having to use all elements of national power to win or outlast the other guy.

However, the mission of 10th Special Forces and our approach to contesting the Soviets evolved during my tenure. By 1982, when I left 10th Special Forces, strategic reconnaissance had largely replaced unconventional warfare as the priority mission in Europe, with Special Forces tapped to be our eyes and ears if and when the Soviets came through the Fulda Gap. The effect of this shift from unconventional warfare to strategic reconnaissance was to recast Special Forces in Europe from being a strategic asset to simply another surveillance system. By the mid-1980s, this would become the primary role for Special Forces in the Army's AirLand Battle concept, providing support to operational-level conventional maneuver.

This was only part of a broader American shift away from irregular warfare occurring at the time. While I was playing a minor role in the development of strategic reconnaissance, as a supporting member of the Flintlock unconventional warfare exercise in Europe, events were unfolding in Iran that would irrevocably change the U.S. special operations community. Operation Eagle Claw's failure in 1980 to recover the American prisoners held in the U.S. Embassy in Iran led to the formation of U.S. Special Operations Command (SOCOM) and a dramatic resourcing shift toward unilateral special operations capabilities. The result would be an unprecedented U.S. capability to strike targets, conduct raids, and execute rescue and recovery operations across the globe.

A consequence of these two major trends was that the strategic thought, preparation, and resourcing required to capitalize on or defend against foreign resistance movements slowed significantly. Specialized capabilities that had been developed to support the unconventional warfare mission were transitioned to these new priority missions, as the Army and the emerging joint force prepared itself for conventional war in Europe and global counterterrorism operations.

Unconventional Warfare in Europe: 10th Special Forces in the Cold War

At Fort Devens, I supported the 10th Special Forces battalion focused on Eastern Europe, which at that time included Bulgaria, Hungary,

Romania, and Yugoslavia.[1] My military intelligence support team was composed of almost entirely Special Forces–qualified intelligence analysts, and our mission was to prepare our battalion's Special Forces teams for unconventional warfare across this region. It was 1979 and the Cold War still dominated U.S. national security, and we were preparing our soldiers to lead the military fight behind Soviet lines.

The 10th Special Forces Group had been established in 1952 to provide the United States an unconventional warfare capability,[2] tasked to "stand-up indigenous resistance forces in the eventuality that Communist forces—namely, the Soviets and the Warsaw Pact—invaded Europe."[3] War on the plains of Europe in the era of thermonuclear weapons was to be avoided at all costs. The consequences of that war were unknowable but also rightly judged by most to be existential on the level of our species. But if war was to break out, 10th Special Forces was to be at the vanguard of efforts to halt its spread, tasked to exploit the discontent and unrest within the populations living under communism, the inherent weakness of totalitarian regimes.

The creation of 10th Special Forces gave the United States the ability to tap into unrest that was still evident in occupied Eastern Europe during the early days of containment. This included all the Eastern Bloc countries, down to and including Yugoslavia, and in the

---

[1]   This was 3rd Battalion, 10th Special Forces. I would later, from 1997 to 1999, have the privilege of commanding this battalion.

[2]   Special Forces were formed in 1952 and led by active duty officers who were former members of the Office of Strategic Services, which had served as the primary special operations and intelligence capability of the United States during World War II. Although many of the Office of Strategic Services capabilities were absorbed into the newly forming CIA, the Army retained much of the office's guerrilla warfare expertise from World War II. The Army, recognizing the value of this expertise as a counter to emerging threats from guerrillas worldwide, formed the Psychological Warfare Division, which by 1952 would establish the "operational group," later referred to as a Special Forces unit, as a construct to "organize existing guerrilla forces or resistance groups capable of conducting strategic and tactical operations against an enemy" (Field Manual 3-18, *Special Forces Operations*, Washington, D.C.: Headquarters, U.S. Department of the Army, May 2014, p. 1-4). For a truly authoritative history of the formation of U.S. Army Special Forces, see Alfred Paddock, *U.S. Army Special Forces: Its Origins*, Washington, D.C.: National Defense University Press, 1982.

[3]   Field Manual 3-18, 2014, p. 1-5.

Soviet Union itself—from the Murmansk Peninsula, at its northwest corner along the Finnish border, and into its depths. These missions envisioned linking up with partisans, presumably known to our intelligence agencies, and then organizing and training them to resist their occupiers or their oppressive regimes.

My military intelligence support team's mission was to help the Special Forces teams as they prepared for and then executed these unconventional warfare missions. In advance of operations, we provided critical information about the physical geography of the operating area, its inhabitants, and enemy military and police forces. We also identified targets for sabotage and interdiction, possible population-control measures, and potential groups (and their leaders) that might offer some resistance potential. During the missions, which would last well over a year and could theoretically last the duration of the war, we would provide the Special Forces teams with emerging information and intelligence as they became available. Strict compartmentation and adherence to operational security regulations were a given and critical to mission success.

We worked closely with 10th Special Forces "area specialist teams." These teams, composed of seasoned Special Forces operators, functioned as the repository of regional knowledge and were critical to the efficacy and survival of our Special Forces teams in the field. Area specialist teams were responsible for preparing a Special Forces team for an operation, drawing on their own operational experience in that area and other available intelligence assets. They would also support the Special Forces team throughout an operation, safeguarding the compartmented information (e.g., identities, escape and evasion plans) on the team's mission.[4] At that time, a team's plan and compromise

---

[4] "The AST [area specialist team], consisting of an area specialist officer and an area specialist [noncommissioned officer], assists in precommitment planning, coordinates activities of their respective detachments in the isolation area, and is responsible for following through on all messages to and from committed detachments. During preinfiltration briefings, a close rapport is established between the alerted detachments and their respective AST. The AST keeps the commander and staff informed of the operational situation" (Field Manual 31-21, *Special Forces Operations*, Washington, D.C.: Headquarters, U.S. Department of the Army, June 1965, p. 38).

code would be known to only that team and its area specialist team, as the risk to a team deployed behind enemy lines was extremely high.

Unconventional warfare, like all irregular warfare, is characterized by a high degree of uncertainty. Together, the military intelligence support and area specialist teams provided the intelligence background and operational experience necessary to reduce the number of unknowns facing our Special Forces teams. Helping our teams survive and succeed as they deployed behind enemy lines shaped nearly every aspect of how we organized ourselves, prepared, and operated. In many cases, there would be no anticipated means of recovery. But our leaders knew the risk, and accepted it, as did the teams and those of us who had signed up for Special Forces.

However, by 1980, strategic reconnaissance began to replace unconventional warfare as the priority mission for Special Forces in Europe (discussed in the next section). In the short term, the deep local and regional knowledge of the area specialist teams remained of great value, and the area specialist officer and the veteran Special Forces soldiers on the officer's teams transitioned to support the new focus on raids and reconnaissance missions. But these missions were of a much shorter duration, and the new focus on collaboration, rather than compartmentation, gradually eroded support for this specialized capability. Not surprisingly, as most big things are cyclical, a requirement for the area specialist team construct reemerged in 2001, when unconventional warfare became one of the leading policy options in the beginning days of our war on terror.

Emergence of Strategic Reconnaissance: The Beginning of Special Forces' Transition to a Supporting Role

In October 1979, I headed to England for my first major exercise with 10th Special Forces and my first trip overseas with the Army. My job was to support scenario development for Flintlock, 10th Special Forces' annual unconventional warfare exercise conducted with key North Atlantic Treaty Organization (NATO) and interagency partners.

Little did I know, but this iteration of Flintlock would prove an inflection point in the role of Special Forces in Europe. At the time, we were simply told to include strategic reconnaissance, alongside unconventional warfare, as a core component of the following year's Flintlock exercise.[5] So, the three of us designing the scenarios (two lieutenants and a captain) dutifully adapted a long-standing Danish commando exercise that would test teams' abilities to identify and report on Soviet troop movement behind enemy lines. Strategic reconnaissance had been a Special Forces mission since at least Vietnam (e.g., Ho Chi Minh trail), but unconventional warfare had always been the priority mission for Europe.

This request to include strategic reconnaissance reflected a major shift in the role of Special Forces in the European theater. The Supreme Allied Commander would later, in 1981, make it clear that he wanted Special Forces to be the eyes and ears for NATO,[6] because if "war should come, he wanted to know when and where the Warsaw Pact units were moving, preferably as it was happening."[7] By 1982, when I left 10th Special Forces and transitioned to VII Corps, strategic reconnaissance had become arguably the primary mission for Special Forces in European war plans. Unconventional warfare would be a follow-on mission, but often only as means for exfiltration, as escape and evasion were the only way that our soldiers would likely get home. Effectively, Special Forces had been recast from a strategic asset to simply another surveillance system in our contest with the Soviets.

---

[5]   This request was from the Special Operations Task Force Europe, which was responsible for the integration of Special Forces and CIA and running a host of exercises for 10th Special Forces (e.g., vetting real-world escape and evasion networks).

[6]   Because the Special Forces at the time were a national asset and held under a national command, the intelligence would actually be provided to U.S. leadership that would then be transmitted to NATO. Clandestine networks were purportedly established throughout Western Europe, "including Belgium, Denmark, France, Germany, Greece, Italy, Luxemburg, Netherlands, Norway, Portugal, Spain, and Turkey, as well as the neutral European countries of Austria, Finland, Sweden, and Switzerland" (Daniele Ganser, "Terrorism in Western Europe: An Approach to NATO's Secret Stay-Behind Armies," *Whitehead Journal of Diplomacy and International Relations*, Winter–Spring 2005, p. 69).

[7]   Stejskal, 2017, p. 196.

In 1979, as we were planning for Flintlock, strategic reconnaissance was in a developmental stage and naturally fell out into two complementary missions. The first was *strategic intelligence collection*, with Special Forces teams building clandestine hide sites along major lines of communication to provide intelligence on the movement of Soviet forces transiting the area. Teams were trained to identify both certain types of materiel (e.g., mobile radar), even if camouflaged or disguised, and the association of that materiel to the Soviet's military structure. The second mission was *target acquisition*, which meant getting that intelligence back to headquarters as quickly as 1980s technology allowed. The intent was that this information would allow NATO to react with long-range indirect fires and aircraft to interdict the movement of critical Soviet assets.

That year's Flintlock exercise combined these two tasks, strategic reconnaissance and target acquisition, into a single requirement that would replace (or at least preface) traditional unconventional warfare on the plains of Central Europe.[8] During the exercise, Special Forces teams would surveil a line of communication (e.g., rail, road) to identify key pieces of equipment and then either interdict the target themselves or simply report it back to headquarters. After completing this primary mission, the teams would rely on the help of willing locals to support the long movement back to friendly territory or begin the hard work of creating or using resistance potential in the enemy's rear area.

The survival of a team following strategic reconnaissance was therefore dependent on the unconventional warfare capabilities of the

---

[8] We even gave this capability a new name, *strategic intelligence collection and target acquisition* (SICTA). Scroll forward a couple of years to my one conventional assignment, which was in Germany as a military intelligence officer with the VII Corps counterintelligence company: Being the only Special Forces–qualified military intelligence guy in the Corps Military Intelligence Brigade, the leadership directed me to participate in a planning session for their support to a "secret" Special Forces exercise called Flintlock. I decided for some reason not to let the Special Forces team know that I was Special Forces–qualified. To them I was just another military intelligence staff officer. Seated next to a newly minted Special Forces officer on staff waiting his turn to take command of a Special Forces detachment, I asked him, "Hey, so what are you guys in group working on?" He leans over and whispers, sharing something that was obviously highly secretive: "Well, we got this thing called SICTA . . ." I about fell out of my chair laughing.

teams. In some cases, we doubted that friendly indigenous contacts even existed. The odds seemed long that someone with a little black suitcase with names and addresses would actually show up in our isolation area. After the Berlin Wall fell and we had access to Eastern Europe and even Russia, it was clear that many of these missions would have been extremely difficult and that there were few known assets to support our missions. Most of the teams, if they survived, would have had to make the long walk back.

In 1984, while participating in a command post exercise at NATO's Allied Forces Central headquarters during my subsequent three-year assignment to VII Corps, I grew to appreciate the importance of this shift to strategic reconnaissance. Among my tasks was the job of helping deconflict allied special operations efforts with deep strikes, some of which may have been with nuclear weapons as Allied Forces Central fought the Warsaw Pact. As I recall, eleven corps-level units, arrayed from the Jutland Peninsula in the north to the Alps in the south, would be involved in operations. The scale of the endeavor was hard to fathom, and it was evident to me that the Special Forces brand of warfare, one marked by infiltration, did not win such wars. We were clearly supporting players, an important enabler perhaps. But such a war would be won by fire and maneuver, by capacity and potential, and by will and Providence.

This shift to strategic reconnaissance, however necessary for the realities of the Cold War, would have long-term repercussions. By the mid-1980s, this mission would become a primary role for Special Forces in the Army's AirLand Battle concept. A consequence of this shift was that strategic thought, preparation, and resourcing required to capitalize on or defend against foreign resistance movements slowed significantly.

Operation Eagle Claw: The American Pivot Toward Counterterrorism

It was during the Flintlock exercise in 1980 that we got word of the failure of Operation Eagle Claw in Iran. At the time, 10th Special

Forces had teams deployed across Europe (from Norway to the Mediterranean) in the blended strategic reconnaissance and unconventional scenarios that we had developed. But we all had friends and teammates that had been reassigned for classified activities, many of whom had been part of the hostage rescue effort in Iran.

For me, the failure of Operation Eagle Claw was my first "our world is about to change" moment. Watching the Armed Forces Networking report on TV, seeing the burned-out wreckage and charred remains, I knew for sure that the business I had chosen a short two years prior was at a crossroads.

What came from the failure was the creation of a hierarchy of headquarters and supporting civilian government agencies to ensure that such a national embarrassment would never happen again. In the ensuing years, that failure would eventually lead to the emergence of a historically unprecedented and peerless counterterrorism capability, one with a truly global reach.

The creation of this world-class counterterrorism capability had two profound, and I believe unintended, effects on America's irregular warfare capability. The first was a major shift in the focus of the special operations community's thinking, which accompanied the creation of Delta Force, the two-star Joint Special Operations Command, the addition of the Rangers to special operations, and the building of the 160th Special Operations Aviation Regiment (out of the helicopter detachments previously assigned to the Special Forces groups). Before, the focus had been on either (1) developing the special operations capabilities that Special Forces, Civil Affairs, and Psychological Operations needed to effectively contest the Soviets in the European theater of the Cold War or (2) providing largely Special Forces advisers to help countries beat back communist insurgents, in such backwaters as El Salvador and Honduras. But as this new global counterterrorism mission gained priority for resources, improving America's ability to "find, fix, and finish" targets on a global scale came to dominate our thinking.

This new approach also cemented a fissure within Army special operations that had emerged in 1977. In the wake of Israeli operations in Uganda and Somalia,[9] President Jimmy Carter first established a requirement for a national counterterrorism capability. Two parallel efforts within the Army to develop this capability emerged. The first was called Blue Light, the "nation's first official antiterrorist team," which 5th Special Forces had established and begun training in advance of the Department of Defense request for this capability.[10] The second, which was modeled on the British Special Air Service and established a direct funding and command relationship with the chief of staff of the Army, was 1st Special Forces Operational Detachment Delta, or Delta Force.[11] Animosity between these two groups emerged almost immediately, and the Army eventually deactivated Blue Light in 1978, following pressure from U.S. national leadership to select a single counterterrorism capability.[12] The resulting struggle created an unhealthy competition in Army special operations that would persist throughout my career.

---

[9] In July 1976, the Israelis executed a counterterrorism raid that captured the attention of the world, rescuing both passengers and crew from an aircraft that had been hijacked by Palestinian terrorists and flown to Entebbe, Uganda. The following year, in October 1977, the German counterterrorism unit Grenzschutzgruppe 9 (GSG-9) liberated both crew and passengers from a hijacked German airplane in Mogadishu. For a discussion of the requirement, see Charles Beckwith and Donald Knox, *Delta Force*, New York: Harcourt Brace Javanovich Publishers, 1983, p. 117. The Department of Defense had reportedly recognized the need to "set up a permanent antiterrorist organization" in the wake of Entebbe (Gary O'Neal and David Fisher, *American Warrior*, New York: Thomas Dunne Books, 2013, p. 142).

[10] O'Neal and Fisher, 2013, p. 143. The authors continue: "[The] mission was to be ready to meet threats from terrorists, to offer individual protection measures, and to learn resistance to interrogation, how to manage hostage situations, surveillance, improvise explosives, kidnapping, vehicle and personal ambush, escape from captivity, and sentry stalking and silent killing."

[11] Beckwith and Knox, 1983.

[12] Beckwith and Knox, 1983.

CHAPTER THREE

Bolivia and the Department of Defense's Entry into the War on Drugs

In January 1987, I signed into 7th Special Forces' Panama-based battalion and immediately took command of Operational Detachment Alpha 774, a 12-man Special Forces team. A midgrade captain, having already commanded a company in Germany,[1] I was eager for the opportunity to command a Special Forces team. I was particularly happy that I would get the opportunity to do so in Latin America, as I was a good Spanish speaker and had spent considerable time in Panama and Venezuela as a kid.[2]

That April, just a few months after my arrival in Panama, during a month-long exercise with the Panamanian military, our team deployed into Bolivia's main coca-growing region to train and advise a rural counternarcotic police force that the United States had established a few years earlier.[3] We would be the first Special Forces team in what

---

[1] That was the VII Corps' Counterintelligence and Interrogation Company.

[2] Accepting the intertheater transfer to Panama, in which I shipped out from Germany to Panama over the Christmas holidays of 1986, was one of two decisions made in my career about which I did not consult my wife. The other was to accept an opportunity to branch transfer to Special Forces when they stood up the branch less than a year later. I regret neither but had to do some hard negotiating after on both.

[3] This Bolivian counternarcotics force was the Unidad Movil Policial para Areas Rurales (Mobile Police Unit for Rural Areas), which we referred to at the time by its abbreviation, UMOPAR. This force was established in 1983 as part of a narcotics control agreement signed between the Bolivian and U.S. governments (U.S. House of Representatives, *Foreign Assistance and Related Programs Appropriations for 1988: Hearings Before a Subcommittee of the Committee of Appropriations*, Washington, D.C.: Government Printing Office, 1987,

was to be a more than decade-long irregular warfare mission for 7th Special Forces in Bolivia.[4] This mission heralded a new application of America's irregular warfare capability, in which we leveraged an indigenous force to bring America's war on drugs to the producers and distributors of narcotics in South America.

Bolivia was the first country in the United States' expanded war on drugs, as well as the first country in which the Department of Defense played a prominent role in U.S. counternarcotic efforts. The first U.S. military operations in Bolivia began in July 1986, with the Army providing air transport and the DEA advising a mission seeking to "locate and destroy cocaine production laboratories" as part of Operation Blast Furnace.[5] The effects from these operations proved extremely short-lived, in part because the Bolivian counternarcotics police force that the United States had established was ineffective and deeply unpopular—so much so that members were forced from their forward base in the "heart of Bolivia's coca-growing region" in the wake of the operation.[6] Stinging from the inefficacy of these efforts, the U.S. ambassador to Bolivia was soon clamoring for an alternative approach.

---

p. 1759). Before our arrival, UMOPAR had partnered primarily with the DEA, though U.S. Army helicopters provided mobility for Operation Blast Furnace, which was a joint DEA-UMOPAR mission (Michael H. Abbott, "The Army and the Drug War: Politics or National Security?" *Parameters*, December 1988).

[4]   Strengthening Bolivia's beleaguered president was already a "top priority" for the United States, as this president had been democratically elected in 1982 following the ouster of a narcotrafficker-supported Bolivian military dictator in 1981 (Sewall H. Menzel, *Fire in the Andes: U. S. Foreign Policy and Cocaine Politics in Bolivia and Peru*, Lanham, Md.: University Press of America, 1997, p. 9). Operations involving an expanded Department of Defense role had initially been considered for Colombia, Bolivia, and Peru, but Bolivia was selected because Bolivia's military capability was the most ill prepared to contest drug traffickers, and the president of Bolivia felt that he was about to lose control of his country to the traffickers (Abbott, 1988, pp. 100–101).

[5]   Abbott, 1988, p. 95; Michael Isikoff, "DEA in Bolivia 'Guerilla Warfare,'" *Washington Post*, January 16, 1989.

[6]   The quote describing Trinidad, Bolivia is from Joel Brinkley, "The Talk of Trinidad; Bolivian Town Resents Drug Glare," *New York Times*, July 25, 1986, but the fact that they were run out of town was something that I learned very quickly once arriving in Bolivia.

My team, which arrived in the wake of the failures of Operation Blast Furnace, was to be the vanguard for the new approach demanded by the ambassador.[7] Our mission was to professionalize this partner force and assist the DEA in putting pressure on the drug networks in Bolivia's main coca-growing area, the Chapare region. Our partner force was ineffective and corrupt, literally everyone in Chapare was involved in the coca trade in some way, and the region was remote and almost inaccessible by vehicle. Success would require working side by side with personnel from the DEA, CIA, Department of State, and Treasury Department. As my team's warrant officer concluded after an initial site survey: "It was the perfect Special Forces mission." Our six-month mission in the Chapare region would be the best education I could have hoped to receive in working with indigenous forces, and it gave me a deep respect and appreciation for those who chose this line of work. I could not have asked for better teammates.

Ultimately, U.S. efforts in Bolivia would fail. In 2008, after a more than 20-year mission and investment of $3 billion, the United States was expelled from Bolivia, and the Bolivian government legalized the production of coca. By 2016, less than a decade after we had left, coca production had fallen by 25 percent, and cocaine interdictions, now unilateral Bolivian operations, had reached an all-time high.[8] Our $3 billion investment might have enabled these successes, but, if so, the United States was not getting credit for it.

Two central observations emerged from my experience in Bolivia. The first is that the United States did not fail because of a lack of capability among the tactical-level formations of the myriad U.S. agencies involved or our ability to work together. Indeed, at the tactical level, a pretty powerful interagency team emerged in Bolivia, and it was the blending of the disparate agencies' authorities and capabilities, civilian and military, that gave us the flexibility that we needed to succeed.

---

[7] The U.S. ambassador's senior military adviser, a Vietnam veteran, had recommended a U.S. Army Special Forces team be given the mission to professionalize UMOPAR.

[8] Mimi Yagoub, "Challenging the Cocaine Figures, Part I: Bolivia," *InSight Crime*, November 16, 2016; United Nations Office on Drugs and Crime, *Bolivia: Monitoreo de Cultivos de Coca 2016*, La Paz, Bolivia, July 2017.

The second observation, however, is that we did not have the models for designing and implementing an effective approach for this inherently population-centric conflict. I saw the consequences of this vividly in 2005, when I returned to my old facility in Bolivia as the one-star commander of SOCSOUTH. While taking the operations briefing, I was amazed at how much had not changed. After nearly 20 years, we and our Bolivian partners were still seizing large quantities of cocaine and arresting both Bolivian and Colombian narcotraffickers but were doing little to understand or address the underlying problems. We were very familiar with the local and regional personalities and political dynamics but did not have the tools necessary to leverage this insight to achieve enduring effects in this population-centric mission. We were also poorly structured to remember past U.S. activities in the region, which was understandable given the relatively short six-month rotation cycle and the lack of an interagency higher headquarters dedicated to accomplishing this mission. It was to be my first campaign in which we fought not to win but simply not to lose. This would become a common theme in subsequent population-centric campaigns in which I was involved.

Building Bolivia's Counternarcotics Capability: Success at the Tactical Level

I deployed with Operational Detachment Alpha 774 into Bolivia's Chapare region in April 1987.[9] Our mission was to train an existing Bolivian counternarcotics police force in jungle light infantry operations; build a base camp to support sustainment, training, and operations; and, more generally, to "make [the team] an effective unit." The final guidance, which was given to us by the senior U.S. military adviser in Bolivia (a foreign area officer who had been an infantry officer in Vietnam), provided us wide discretion that we would make good use of when the magnitude of the task became apparent.

---

[9]   Our team was augmented by a few additional Special Forces engineers for the mission.

This police force, the Unidad Movil Policial para Areas Rurales (UMOPAR), was established in 1983 by the DEA and the Department of State's Bureau of International Narcotics and Law Enforcement to police Bolivia's Chapare region. Composed of mostly Quechua indigenous people from high in the Andean Ridge, this U.S.-funded unit had a very inauspicious beginning. Its inefficacy was demonstrated most publicly in the failures of Operation Blast Furnace the previous summer, with observers describing members as "lethargic, unimaginative, and less than successful."[10] Further, in the wake of this failure and in spite of the equipment and training they had received from the United States by that point, this Bolivian force was expelled, while on an operation, from the town of Trinidad by the locals at the urging of local drug runners. It was these failures that led to the request from the ambassador, on the advice of his military advisers, for assistance from U.S. Army Special Forces. We would ensure that our Bolivian charges would not get run out of another town.

What became immediately evident was that a sort of informal quid pro quo had developed between all parties involved, with basically everyone benefiting from the status quo. We called this the "Chapare equilibrium." Drug-dealing networks would give new local commanders a "prize" early in their tours to demonstrate to their higher-ups that they, unlike their predecessors, were effective and above reproach. This prize could be a load of drugs or, more often, a successful airplane interdiction. The U.S. civilian elements would draw per diems, get hazard duty pay, spend a month or so in the field, and then return home with a little money and a nice block check for having deployed—there were even rumors of U.S. staff receiving payments from the drug networks.[11] In the meantime, senior Bolivian officials got rich, either running their own networks or extorting protection money from those run by others, and U.S. officials touted the occasional success as proof of progress. Crop-substitution projects became the congressional overseas destination of choice.

---

[10] Menzel, 1997, p. 25.

[11] In 1987, there was an investigation into whether "corruption within the Country Team in La Paz" contributed to the inefficacy of UMOPAR (Menzel, 1997, p. 25).

Our job, as we saw it, was to be a constant disruption to this equilibrium while first building a capable force and later advising it on counterdrug operations. Although we realized that we were not going to win the war on drugs, we became determined to upset the status quo. As the military guys in the mix, we saw ourselves as the fly in the ointment.

Nearly every agency of the U.S. government had a hand in the war on drugs, and our Special Forces teams worked at the tactical level with a variety of what I like to call "civilian special operators" from the U.S. diplomatic, development, and intelligence communities. Our DEA partners, who lived with our team in the early days of the program, brought a synergy of law enforcement and special operations training that was essential to our mission. Although most were volunteers from DEA offices back in the United States, they became pretty good at applying special operations techniques following multiple deployments into Bolivia, and eventually Peru and Colombia, and after they started receiving training at the U.S. Army Ranger school. The border patrol's Border Patrol Tactical Unit brought a tactical proficiency for tracking drug operations and trained our partners in "traffic and transportation checks, vehicle stops, and ground control."[12]

My team brought the skills to survive in dangerous situations and patrol in the jungles, expertise in small-unit military operations, intelligence tradecraft, and an ability to train and build trust with indigenous counterparts. The Special Forces team made the most of training patrols to assist our DEA partners, but we were not authorized to accompany either the DEA or its Bolivian units on operations. What resulted was an early form of remote advise and assist, a mission that would be formalized years later by Special Forces units supporting the Syrian resistance.

We developed a reasonably powerful interagency team, and the blend of each of our unique authorities and capabilities gave us a lot of needed flexibility in achieving our mission. The guidance we had

---

[12] U.S. House of Representatives, Committee on Foreign Affairs, *Report of a Staff Study Mission to Peru, Bolivia, Colombia, and Mexico, November 19 to December 18, 1988*, Washington, D.C.: Government Printing Office, February 1989, p. 19.

received in advance of our operations was to know your left and right limits, know your rules of engagement, and use every single bit of that latitude that you are given to succeed. This interagency team gave us exactly the latitude we needed.

The transformation of our Bolivian partner force over our six-month mission was remarkable. When we arrived, the force had incompetent leadership and members were dispirited, untrained, and unable to plan or execute even the simplest movement in the jungle, competently raid a house, or disable a dirt airstrip. Our partners made what I saw as significant progress by the end of our six months. They reorganized themselves based on our recommendations into efficient and sustainable paramilitary police units, shed unnecessary (and mostly corrupt) officer overhead, and established a Bolivian-run training course that allowed them to professionalize their force on their own. By the time my team left, they had become proficient in transforming newly assigned unit members from street cops into rural drug paramilitary operators, no small feat. There were also dramatic improvements in the physical conditions at their bases,[13] which was doing much for the unit's overall health and morale. Subsequent Special Forces trainers would settle into the background after working themselves out of a job, transitioning into an advisory role that would last for more than a decade.

The improvement in our partners' confidence and morale was evident, in some cases, even after only the first few months of our partnership. The most memorable example of their ingenuity was the formation of a mounted unit within the counternarcotics force that used confiscated dirt bikes to run the many trails over which coca paste and precursory material were moved. Established by one of my Special Forces soldiers, who was an avid motorcycle rider, the "Bolivian CHiPS" (as we called them, after the California Highway Patrol)

---

[13] When we arrived, the main base camp had an open latrine some 30 feet from the well for drinking water. People were living in either tents long past their service life or dilapidated buildings with no water, windows, or beds. By the time we left, each police officer had a clean, dry bunk space in a refurbished building; there was running water; and the latrines were connected to a makeshift but effective sewage septic system. The septic system was a true masterpiece, built by Operational Detachment Alpha 774's exceptional team sergeant.

proved highly effective as a result of their adaptiveness and initiative. The relatively junior-level officers in this unit became very loyal to their American trainers and advisers and became a critical source of intelligence on what was happening inside our partner force and throughout the Chapare region.

By the time we left, the only real remaining challenge was the endemic corruption throughout the force, which was something that we had struggled to address throughout our tenure. We learned about this corruption because of recruited assets from within the unit, which I judged necessary as much for our own force protection as for intelligence on unit corruption and the drug network activity. This unconventional warfare approach, though ordinarily high risk, seemed appropriate given the pervasiveness of the corruption and the potential threat to the team once the unit's effectiveness began to improve.

The commanders and their officers knew that we knew about their corruption but did not care—from their perspective, the drug problem was a U.S. problem and not a Bolivian one. The reality was that our Bolivian counterparts were poor, and drug traffickers compensated them well for simply looking the other way as drugs left their country.

This was openly explained to me at the end of my tour by a Bolivian police colonel, one of the unit's leaders who had worked with me from the beginning of the mission, and become a good friend along the way, and who was for the most part an honest man.[14] One evening after training, he told me that his troops, who were paid about $20 a month, could make double that amount in a single day from petty bribes manning the checkpoint on the one road that traversed the Chapare. He could make $20,000 himself for simply not patrolling in an area that the drug networks wanted to use for a few days, either as a collection center for coca paste or as an airfield from which to carry the paste to waiting labs in Colombia. He asked me plainly: "We are criticized by you Americans for being corrupt, but how does a leader keep that much money from corrupting his officers?" I recall trying to tug at

---

[14] I say "for the most part" pretty much because I was later told that he was arrested on corruption charges, although the charges could easily have been to silence an honest man.

the immorality of corruption, his patriotic duty to his country, and his responsibility as a commander, but it was thin gruel considering that he was outgunned by a ruthless adversary who was often supported by corrupt officials above him. But in Special Forces, your job is to work with what you've got and make it work.

Unprepared for This New Form of Population-Centric Warfare: Difficulties Above the Tactical Level

We were the vanguard of what would prove to be a robust role for the Department of Defense in the U.S. war on drugs. This effort had been ongoing since 1969,[15] but previous civilian-led efforts—from President Richard Nixon's ill-fated effort to control trade along the U.S.-Mexico border to national-level eradication-focused partnerships with "drug source countries"[16]—had not resulted in significant reductions in the supply of drugs inside the United States. By the early 1980s, as a result

---

[15] In March 1969, Nixon followed up on his campaign promise to "move against the source of drugs" and "accelerate the tools and weapons to detect narcotics in transit" by establishing the interagency Special Presidential Task Force Relating to Narcotics, Marijuana and Dangerous Drugs tasked to "formulate a plan for positive and effective action to control the illicit trafficking of drugs across the Mexican border" (Richard B. Craig, "Operation Intercept: The International Politics of Pressure," *Review of Politics*, Vol. 42, No. 4, 1980, p. 556; Special Presidential Task Force Relating to Narcotics, Marijuana and Dangerous Drugs, *Task Force Report: Narcotics, Marijuana and Dangerous Drugs, Findings and Recommendations*, June 6, 1969).

[16] The de facto blockade of the U.S.-Mexico border was implemented under the auspices of Nixon's Operation Intercept, which did reduce the flow of marijuana but at a massive cost to both taxpayers and U.S.-Mexico relations. Senator Barry Goldwater, as an example, reported that "Operation Intercept is an example of how bureaucrats and legislators without vision can destroy so many years of effort on behalf of extremely cordial interamerican relations." Operation Intercept proved short-lived as a result of these challenges, terminating after only some three weeks, and Nixon ultimately issued a formal apology to the Mexicans for the disruption it had caused (Craig, 1980, pp. 559–560, 566, 571, 578). The national-level partnerships involved the United States providing assistance to support crop eradication efforts and law enforcement capabilities in Bolivia, Colombia, Mexico, and Peru (Clare Ribando Seelke, Liana Sun Wyler, June S. Beittel, and Mark P. Sullivan, *Latin America and the Caribbean: Illicit Drug Trafficking and U.S. Counterdrug Programs*, Washington, D.C.: Congressional Research Service, March 19, 2012, p. 9).

of these difficulties, Congress was pushing for an expanded role for the Department of Defense, which until then had only been in a supporting role.[17] Although there was initially resistance from within the Department of Defense for this expanded role,[18] this changed in April 1986, when President Ronald Reagan formally directed the Department of Defense to prepare for and, ultimately, play a key role in the counternarcotics mission with his *Narcotics and National Security* directive.[19]

Three months later, the first U.S. military operations in this new war on drugs would begin in Bolivia, and the Department of Defense

---

[17] Congressional support for an expanded Department of Defense role grew through the 1970s and into the early 1980s amid growing concerns that "law enforcement personnel were ill-equipped to effectively combat well-armed drug cartels and operate in conflict situations in drug source countries" (Seelke et al., 2012, p. 9). And, in 1981, Congress removed what was reportedly the primary impediment to military support to the war on drugs, when it amended the Posse Comitatus Act of 1878 (18 U.S.C. 1385) in the 1982 Department of Defense Authorization Act (Pub. L. 97-86, Department of Defense Authorization Act, 1982, December 1, 1981; U.S. General Accountability Office, *Federal Drug Interdiction Efforts Need Strong Central Oversight*, Washington, D.C., June 13, 1983, p. 74). Previous support to the war on drugs had involved the Department of Defense providing training, equipment, transport, and personnel (e.g., a "radar picket" along the U.S.-Mexico border during Operation Intercept) to civilian agencies (U.S. House of Representatives, statement of Ronald F. Lauve, Senior Associate Director, General Government Division, *Hearing Before the Subcommittee on Crime, House Committee on the Judiciary on Narcotics Enforcement Policy*, Washington, D.C., 1981; Elaine Shannon, *Desperados: Latin Drug Lords, U.S. Lawmen and the War America Can't Win*, New York: Viking, 1988, pp. 48–49).

[18] Although the amendment of the Posse Comitatus Act clearly authorized the sharing of intelligence, equipment, and facilities between the U.S. military and civilian agencies, the Department of Defense initially resisted fully supporting the drug mission, contending that it would affect military readiness (Abbott, 1988, p. 100).

[19] Although Reagan had voiced his support for the war on drugs in 1982 (Ronald Reagan, "Remarks on Signing Executive Order 12368, Concerning Federal Drug Abuse Policy Functions," June 24, 1982), his *Narcotics and National Security* directive made it the "policy of the United States, in cooperation with other nations, to halt the production and flow of illicit narcotics, reduce the ability of insurgent and terrorist groups to use drug trafficking to support their activities, and strengthen the ability of individual governments to confront and defeat this threat" (National Security Decision Directive 221, *Narcotics and National Security*, Washington, D.C.: White House, April 8, 1986). This directive tasked the Department of Defense to provide military forces in support of U.S. civilian-led counternarcotics efforts if that support was requested by the host government (Abbott, 1988, p. 100).

would find itself ill prepared for what would be a population-centric conflict. Our conventional forces were, as the commander of U.S. Southern Command (SOUTHCOM), General John R. Galvin, concluded in 1986, focused "almost exclusively on . . . a massive high-intensity war in Western Europe" and not prepared to fight the "people's war" that "erased the line between military and civilian, between war and politics, between combatant and non-combatant."[20]

Because the Department of Defense lacked the doctrine, knowledge, and training for success in this type of conflict, early operations (e.g., Operation Blast Furnace) were, perhaps unsurprisingly, ineffective.[21] So the department eventually turned to U.S. Special Forces. The counternarcotics mission would showcase Special Forces' inherent capability to deploy into uncertainty, assess the situation, and mitigate problems by working with local actors and among local populations.

However, though successful at the tactical level in our assigned mission, our successes did little to advance U.S. strategic objectives in Bolivia. This first became clear to me when Major General Bernie Loeffke, the U.S. Army South commander during my time in Bolivia and a former Special Forces officer, shared his skepticism of our effort during a visit to our base. He saw the effects we were achieving and agreed that we were doing what we needed to but was skeptical of the Department of Defense in what was an inherently unwinnable mission.

History would prove Major General Loeffke right, as our tactical successes would ultimately damage our long-term goals. The interdiction efforts of my own team, those that came after us, and those of the DEA's Operation Snowcap that followed in later months reportedly created "previously nonexistent resentment and hostility among the

---

[20] John R. Galvin, "Uncomfortable Wars: Toward a New Paradigm," *Parameters*, Vol. 16, No. 4, 1986, p. 5. The third quote in this sentence was quoted by Galvin in Douglas Pike, "Conduct of the War: Strategic Factors," in John Schlight, ed., *The Second Indochina War: Proceedings of a Symposium Held at Airlie, Virginia, 7–9 November 1984*, Washington, D.C.: U.S. Army Center of Military History, 1986, pp. 101–102.

[21] Abbott, 1988, p. 95; Isikoff, 1989.

coca farmers in the Chapare region" and set the "conditions that could lead to future insurgent activities."[22]

I returned to Bolivia 18 years later, in 2005, and had the chance to visit my old facility. The improvements at my old facility were amazing, long gone were our standard-issue Army tents and the obstacle course that we cut from the jungle, and there was now a paved airfield and even hot showers in the visitors' quarters, a far cry from the river bathing that had been the only option 18 years earlier![23] Not long into the courtesy operations update, it became evident that, in spite of the vastly improved physical infrastructure, the reality on the ground was remarkably unchanged from when our team had first started the mission. There were the same bad places and *campesinos*, no doubt many the children of those we had encountered nearly two decades earlier making coca paste, and the sense that the Chapare equilibrium was still at play.

The commander of the same Bolivian counternarcotics force, whose starched fatigues and shined boots told me he'd learned something at the School of the Americas, was briefing operations in the same places we had conducted operations years before. Improvements to infrastructure and even training had not led to accomplishing the mission of stopping the drug trade out of the area. The amount of money involved was too great.

After more than two decades of training, our partner force was still doing little more than "mowing the grass." They had proven themselves an effective law enforcement strike force, capable of seizing—albeit with support from U.S. intelligence agencies—large quantities of cocaine and arresting both Bolivian and Colombian narcotraffickers.[24]

---

[22] U.S. House of Representatives, Committee on Government Operations, *Stopping the Flood of Cocaine with Operation Snowcap: Is It Working?* Washington, D.C.: Government Printing Office, 1990, p. 52.

[23] Our sewage system was about the only thing that remained from our era. My team sergeant, who I am convinced could build anything, would have been proud.

[24] In 1992, an estimated 30 percent of Bolivia's total cocaine production was either interdicted or eradicated because of the efforts of UMOPAR and the eradication-focused Coca Reduction Directorate, and operations during this time—which were part of Operation Ghost Zone—dismantled several important narcotrafficking organizations. The U.S.

But alternative development programs were far too limited,[25] and the economic realities meant that coca production would continue. And their sometimes-heavy-handed approach against coca farmers, which had only been a nuisance in the 1980s while I was there,[26] would contribute to the eventual failure of our war on drugs in Bolivia.

Strategic Failure in the U.S. Mission in Bolivia: Inadequacy of Existing Models and Structures for This Population-Centric Conflict

U.S. efforts in Bolivia were not effective in reducing cocaine production. During the 1980s, the Bolivian cocaine industry thrived, growing an average of 35 percent per annum,[27] despite nearly $600 million in U.S. taxpayer support to the Bolivians and our partnerships across a multitude of security forces.[28] Political resistance to our approach would play a role in the election of President Evo Morales,[29] who

---

ambassador was paraphrased as saying that this success meant that "narcotraffickers had been prevented from gaining 'meaningful influence' in the political process and . . . the Bolivian government [was] able and willing to assert its authority" (Menzel, 1997, pp. 73–74 and 92).

[25] William W. Mendel, "Illusive Victory: From Blast Furnace to Green Sweep," *Military Review*, December 1992, pp. 79–80.

[26] UMOPAR and the DEA were frequently accused of human rights abuses, with one Bolivian group alleging that the DEA-directed "UMOPAR routinely attacked coca farmers, stealing money, goods and other personal property." This frequently led to organized resistance from the coca farmers, both locally and in national-level politics (quoted in Menzel, 1997, p. 33).

[27] Estimates are as reported in Menzel, 1997, p. 2.

[28] These data are from World Bank, World Development Indicators, data set, accessed 2018 ("Net bilateral aid flows from DAC donors, United States [current US$]"). The 1987 Principles of Narcotics Cooperation provided $300 million during just the 1987–1989 time frame (Menzel, 1997, pp. 101 and 103).

[29] Linda Farthing and Benjamin Kohl, "Social Control: Bolivia's New Approach to Coca Reduction," *Latin American Perspectives*, Vol. 37, No. 4, 2010, p. 198.

expelled both the DEA and U.S. ambassador in 2008 for purported political meddling.[30]

Contemporary observers highlighted a lack of U.S. government coordination as a key impediment to our efforts in Bolivia and in the war on drugs more broadly.[31] At the operational level, and despite the Department of State being assigned as the lead agency for in-country coordination, there was reportedly "bureaucratic competition" as different U.S. agencies competed for the "counternarcotics pie."[32] A lack of clear national-level policy guidance also led, reportedly, to poor integration of U.S. efforts with that of the host government, so that the programs did not reflect "the priorities of the host governments we are attempting to assist."[33]

One proposal for overcoming the challenges encountered in the war on drugs concluded that U.S. government coordination must be at the highest national level and that "any governmental level below the National Security Council (NSC) will not be able to compel major U.S. agency players to get their houses in order."[34] And though the Department of State must be in the lead, it must be "forced to rise to competent stewardship" and effectively integrate the many elements of U.S. power that are necessary for success in these types of endeavors.[35]

There were similarly no models or doctrine for designing and implementing the President-mandated whole-of-government approach. At its core, our approach did not reflect a simple fact—that coca production employed 30 percent of the Bolivian labor force and was by far the most important export commodity.[36] Thus, although the alter-

---

[30] Clare Ribando Seelke, *Bolivia: In Brief*, Washington, D.C.: Congressional Research Service, 2014, p. 6.

[31] August G. Jannarone and Ray E. Stratton, "Toward an Integrated United States Strategy for Counternarcotics and Counterinsurgency," *DISAM Journal*, Winter 1990–1991.

[32] This observation is attributed to a "senior official of Project Alliance" in Jannarone and Stratton, 1990–1991, p. 54.

[33] Jannarone and Stratton, 1990–1991, p. 54.

[34] Jannarone and Stratton, 1990–1991, p. 55.

[35] Jannarone and Stratton, 1990–1991, p. 57.

[36] Menzel, 1997, p. 2.

native development assistance programs at the core of our eradication strategy were robust, they were far from sufficient for managing the task at hand.[37] And it is likely that our alienation of the local population and host government, by using an endemically corrupt force and not restricting our operations to only narcotraffickers, endangered U.S. strategic interests in Bolivia.

Although our war on drugs in Bolivia might have proved ill-fated, there is evidence of a steady improvement in the professionalism and training of the Bolivian police. Looking back on our Bolivian experience, I think that there was a "values transfer" from adviser to the advised even if we failed in achieving the intended U.S. strategic objectives, a phenomenon that our predecessors saw in Vietnam and I would see in El Salvador, Colombia, Afghanistan, and Iraq in the years to come. In early 2020, the Bolivian military attaché sought me out through friends at the National Defense University. It seems that the new Bolivian government wanted to recognize our team for the assistance rendered more than three decades ago. Perhaps our time horizon to gauge success was too short.

---

[37] One contemporaneous estimate suggests that a rough tenfold increase in resources would have been necessary to support the desired end state of coca eradication (James A. Inciardi, *Handbook of Drug Control in the United States*, New York: Greenwood Press, 1990, p. 212).

CHAPTER FOUR

El Salvador and the Fight Against Communism in the Americas

El Salvador was my first taste of counterinsurgency fighting, and I would travel there several times during 1988 and 1989 as part of U.S. efforts to contest a Soviet-supported insurgency. By this time, I was responsible for six Special Forces teams, having taken command of Alpha Company in the Panama-based 3rd Battalion of 7th Special Forces, and assisting the U.S. Embassy in El Salvador with improving the warfighting skills of the Armed Forces of El Salvador would be one of our core missions.

U.S. military support to El Salvador had begun in 1981, as a response to irrefutable evidence that the Soviet Union was supporting a Salvadoran insurgency.[1] With Congress fearful of another Vietnam and the Reagan administration's attention turned to imposing costs directly on the Soviet Union, the U.S. military commitment to El Salvador was deliberately limited. The Carter administration initially committed only 19 U.S. military advisers,[2] and Reagan increased this number only modestly—to a maximum of 55 uniformed military

---

[1] In 1981, the newly elected President Reagan's secretary of state, Alexander Haig, declared that the United States would "not remain passive in the face of this Communist challenge, a systematic, well-financed, sophisticated effort to impose a Communist regime in Central America" ("Excerpts from Haig's Briefing about El Salvador," *New York Times*, February 21, 1981).

[2] Attributed to Americas Watch Committee and American Civil Liberties Union, *Report on Human Rights in El Salvador*, New York: Vintage, 1982, p. 189, in Paul Cale, "The United States Military Advisory Group in El Salvador, 1979–1992," *Small Wars Journal*, 1996.

advisers.[3] These military advisers were supplemented by small teams of Special Forces soldiers from our battalion, with wide-ranging duties that included helping with basic training, advisory work with combat units, and working with U.S. development and intelligence agencies.

Supporting U.S. efforts in El Salvador had been our battalion's primary mission for nearly five years. Our battalion had roughly 40 to 50 Special Forces operators, the equivalent of three to four Special Force steams, in El Salvador on six-month orders at almost all times. The burden of this mission would fall to just two of our battalion's companies, as the third company was restricted to its counterterrorism mission. The El Salvador mission thus used roughly a third of our force strength at any given time. When the six-month counternarcotics mission in Bolivia was added in 1987, and Peru the following year, it meant that Alpha and Bravo Companies were emptied out for half the year on real-world missions. And the teams deployed on shorter (one- to two-month) deployments for training to other countries in the region for the other six months. It was a fast-paced and exhilarating battle rhythm that would become familiar to those in Special Forces after September 11.

Commanding Alpha Company, known among ourselves as the Alpha Gators, would prove a formative experience. I had come out on the promotion list to major in early 1988 and, as the senior promotable captain, the battalion commander tagged me to command the company. One day I was just one of the company's six team leaders, and the next day I was the boss. I would long regret my abbreviated time on a Special Forces team, but commanding a company at the vanguard of both our counterdrug and counterinsurgency missions was an ideal

---

[3]   And Reagan was explicit that these advisers would not accompany El Salvadoran forces into combat:

> Now, you use the term "military advisers." You know, there's sort of technicality there. You could say they are advisers in that they're training, but when it's used as "adviser," that means military men who go in and accompany the forces into combat, advise on strategy and tactics. We have no one of that kind. We're sending and have sent teams down there to train. They do not accompany them into combat. They train recruits in the garrison area. (President Ronald Reagan, "Excerpts from an Interview with Walter Cronkite of CBS News," March 3, 1981)

place to learn the foreign internal defense side of the Special Forces profession.[4] Taking command of a company that had been recently led by two Special Forces' legends, Major Roy Trumble and Sergeant Major Kenny McMullan, meant that there was a very high bar for excellence in command.[5] It was my everlasting good fortune that Kenny would remain the company sergeant major throughout my command tour.

U.S. support to the El Salvadoran government would take "far longer" (11 years) and "cost far more" (some $6 billion) than anyone expected,[6] but was ultimately successful.[7] The United States and its Salvadoran allies did not defeat the Farabundo Martí National Liberation Front communist insurgency in the classical sense. But the El Salvadoran military facilitated a peaceful transition to democratic civilian control in 1992, which was almost certainly a result of deliberate efforts by the United States to shape the mind-set and composition of the military's leadership.[8]

Our success in El Salvador was at least partially accidental. For one, our operational approach did not reflect a complete understanding of how to contest this counterinsurgency. Our enduring mission was

---

[4]   El Salvador, truth be told, by this time was a routine, albeit exciting, mission for the battalion. Most six-month missions in El Salvador were not whole-team missions, except La Union at its basic training location, which I think made the counternarcotics missions in Bolivia and Peru in some ways more desirable. It helped too that the early counternarcotics missions had few rules of engagement.

[5]   Kenny Mac was a Son Tay raider, a legend in the Special Forces community, and as it turns out one of the smartest men with whom I would serve. Major Trumble would return to command 3rd Battalion within a few months of my taking command of his old company, and over the next year and a half, I had the privilege to both command and serve as his operations officer in combat. He was to be the finest commander I would have in my 37 years on active duty.

[6]   Benjamin C. Schwarz, *American Counterinsurgency Doctrine and El Salvador: The Frustrations of Reform and the Illusions of Nation Building*, Santa Monica, Calif.: RAND Corporation, R-4042-USDP, 1991, p. v.

[7]   Some international groups purportedly treated the outcome as a success for the Farabundo Martí National Liberation Front, since it was not defeated (Victor M. Rosello, "Lessons from El Salvador," *Parameters*, Winter 1993, p. 107).

[8]   Michael J. Hennelly, "US Policy in El Salvador: Creating Beauty or the Beast?" *Parameters*, Spring 1993.

to help transition an ineffective, passive, largely conscript, and conventionally minded El Salvadoran force into an aggressive and effective counterinsurgency force. In this, we were, in my judgment, at best marginally successful. Selected elite units and conventional rapid reaction battalions performed well enough, but it remains an open question whether they meaningfully threatened the insurgency. The greatest benefit of our efforts, it would later turn out, was that our presence reduced abuses by El Salvadoran units that had given the insurgency so much potency.

The limitations on the size of our presence, the 55-person "speed limit" that was an artifact of the then-recent experience of the United States in Vietnam, also proved an unexpected boon. The restriction held off pressure from many in the U.S. security sector to fall back on a classic strategy of mass and firepower in pursuit of a decisive victory. Instead, by supporting the El Salvadoran military only to an extent that would prevent defeat, we gave the El Salvadoran government an essential element found in every successful counterinsurgency campaign: time. Decisive victory against the insurgency, which as we would learn later in Afghanistan and Iraq, is nearly impossible under today's international rules. We gave El Salvador the time needed for political reform that allowed for the eventual success of the peace settlement.[9] This limitation forced on us the strategic patience that is often essential in irregular warfare.

Building an El Salvadoran Counterinsurgency Capability: Provisional Success at the Tactical Level

The central goal of U.S. military assistance to El Salvador was to build capable counterinsurgents within the Armed Forces of El Salvador, as the Salvadorans had struggled to make progress against their

---

[9] See Ernest Evans, "El Salvador's Lessons for Future U.S. Interventions," *World Affairs*, Vol. 160, No. 1, 1997, for a discussion of the importance of democratization.

domestic insurgency.[10] The hope was that this military assistance might help shift the Salvadorans away from an "ineffective, conventional warfighting strategy of indiscriminate air attacks and undisciplined sweeps through guerilla-controlled territories" to a strategy that combined "small, lightly armed units, pinpointed operations assisted by 'hunter-killer' squads . . . [and] civil defense units, regarded as an indispensable aspect of counterinsurgency warfare."[11] This assistance was part of a broader effort by both the Carter and Reagan administrations to bolster a Salvadoran civil-military government that had forcibly seized power in October 1979 in the wake of an analogous insurgency that had toppled the Nicaraguan government.[12] This new Salvadoran government, like us, viewed communism as an existential threat to their way of life.

The 7th Special Forces Group was a central component of these efforts, providing the primary U.S. manpower for the training of the El Salvadorans. Our Special Forces soldiers joined a cadre of Military Advisory Group advisers (most of whom were themselves Special Forces officers and many with tours in 3rd Battalion, 7th Special Forces) who were there on one-year assignments to the El Salvador brigade headquarters, civil defense organizations, and basic training school in La Union.

While the Military Advisory Group advisers focused on building the institutional capacity of the Armed Forces of El Salvador, it was our battalion's responsibility to improve their counterinsurgency proficiency. Our soldiers would deploy on both six-month temporary duty tours and shorter deployments of six to 12 weeks. These shorter deployments were critical because they did not apply against the 55-person cap, and we would frequently deploy individual teams or even single

---

[10] The Salvadorans had reportedly lacked "the doctrine, structure, ideology, and mentality to fight a counterinsurgency war" (Schwarz, 1991, p. 17).

[11] Schwarz, 1991, p. 17.

[12] They removed a deeply unpopular president "in hopes of forestalling" mass insurrection by insurgent and civic groups that had toppled the Nicaraguan government in July 1979 (Mark Peceny and William D. Stanley, "Counterinsurgency in El Salvador," *Politics and Society*, Vol. 38, No. 1, 2010).

soldiers on short tours to make up for the lack of U.S. manpower for training. This allowed us to have as many as 150 U.S. advisers, albeit still a very limited number, in theater at any given time.[13]

The type of training and the approach to implementing that training varied with the unit. For El Salvador's conventional units, training was conducted at their bases and focused on developing their counterinsurgency capabilities, which included such skill areas as communication, intelligence, and logistics.[14] Rather than one-off training sessions, El Salvador's special operations units received nearly continuous training and operational support from U.S. special operations personnel and their interagency partners. This training was nearly as advanced as that of U.S. special operations and included counterinsurgency, interdiction, and crisis response capabilities.

This training was reportedly quite successful. One observer concluded that "the use of Special Forces advisors at the small unit level proved to be the best use of the 'train the trainer' technique in attempting to solve the long term problem of creating an effective" Armed Forces of El Salvador.[15] The U.S. Ambassador to El Salvador Thomas Pickering similarly lauded the value of Special Forces in developing the civil defense program.[16]

However, despite this success, the El Salvadoran conventional units were strikingly weak, almost shockingly so, given the effort we had invested in them. I saw this firsthand during a three-week deployment to train a few squads from one of El Salvador's immediate reac-

---

[13] The 55 advisers on one-year orders consisted of Special Forces–qualified advisers at the Armed Forces of El Salvador brigades, the military group staff at the U.S. Embassy, helicopter training and maintenance personnel, planning teams, and small training teams (five to six personnel) focused on supporting Armed Forces of El Salvador recruitment, the El Salvadoran Navy, or providing garrison training (Cale, 1996, pp. 14–15).

[14] Attributed to Tommie Sue Montgomery, *Revolution in El Salvador: Origins and Evolution*, Boulder, Colo.: Westview Press, 1982, in Cale, 1996, fn. 24.

[15] Cale, 1996.

[16] Max G. Manwaring and Court Prisk, *El Salvador at War: An Oral History from the 1979 Insurrection to the Present*, Washington, D.C.: National Defense University Press, 1988.

tion battalions that specialized in guerrilla warfare.[17] This unit was one of the El Salvadorans' few offensive tools, designed to keep pressure on the insurgents.[18] Despite the fact that the squads were pulled directly out of combat to participate in our assessment and training, we found ourselves limited to training only the most basic of skills (e.g., zeroing weapons, basic marksmanship).[19] Looking back, it is possible that the unit's commander was simply trying to placate the Americans and sent his worst squads. But the result was that, because of our short mission, we had little time to do much beyond that.

This less-than-positive experience came after nearly six years of advisory work by 7th Special Forces in El Salvador and a significant U.S. investment in the institutional training and professional development programs of the Armed Forces of El Salvador. From my foxhole, this challenge pointed to the importance of time in a population-centric fight, the need to keep expectations realistic, the inherent weakness of a conscript army, and the need to, regardless, make the best of indigenous solutions.

In 1989, during my last trip to El Salvador, it became clear to me that we had built a counterinsurgency capability that was effective at the tactical level but still struggled at designing and executing company-level operations without our support, a conclusion reached

---

[17] This was the Batallón de Infantería de Reacción Immediata Bracamonte (BIRI Bracamonte) of the El Salvadoran Army's 1st Brigade. For a background on this unit, see Centro Documental Historico Militar, "Batallon Bracamonte," webpage, undated.

[18] The rest of 1st Brigade was largely doing infrastructure protection. My experience would be that this division of labor, between conventional units and elite infantry, commando or special operations would play out repeatedly in counterinsurgency efforts elsewhere.

[19] The poor state of training pointed to a chronic weakness of many of the militaries I would work with—specifically, their lack of a capable noncommissioned officer corps. They would improve after years of coaching, teaching, and mentoring by U.S. advisers and their increased participation in the U.S. Army's noncommissioned officer education system, initially in the United States, then in their own similar courses. An El Salvador noncommissioned officer corps, with roles equal to that of the U.S. Army, would take years and require patient and long-term investment. The state of training also meant accepting the fact that junior officers would perform the tasks typically given to noncommissioned officers. After years of tenacious effort by U.S. noncommissioned officers, El Salvador and Colombia would both build reasonably strong noncommissioned officer corps.

by other contemporaneous observers.[20] This difficulty would characterize subsequent U.S. counterinsurgency efforts as well. Advise and assist missions at the unit level are relatively straightforward, as failure to learn means an increased likelihood of death, a fact that focuses the mind.[21]

Professionalizing the Armed Forces of El Salvador: Success at the Operational Level

The U.S. mission in El Salvador was an eventual success,[22] and our military efforts almost unquestionably contributed to that success. However, while our military mission was focused on improving the lethality of El Salvador's counterinsurgents, the greatest benefit of U.S. training and partnering was the "institutional conversion to a professional military and the dramatic improvement of its human rights record."[23]

This partnering transitioned the "corrupt, barracks-bound Salvadoran military whose only significant victories were against the civilian population" to an organization that was both competent and had improved respect for human rights.[24] It was reportedly the "day-to-day

---

[20] Attributed to Andrew J. Bacevich, James D. Hallums, Richard H. White, and Thomas F. Young, *American Military Policy in Small Wars: The Case of El Salvador*, Washington, D.C.: Pergamon-Brassey's, 1988, in Bobby Ray Pinkston, *The Military Instrument of Power in Small Wars*, Fort Leavenworth, Kan.: U.S. Command and General Staff College, 1996.

[21] Culture, history, education level, corruption, resource, and technology constraints, as well as philosophical differences about human life and war and peace, add to the challenge of the adviser and the advised. The more distinct the differences between the countries giving and receiving the assistance, the more difficult the advisory mission. The further the advised function is from the reality of the battlefield, and the higher up in the chain of command it is, the more difficult it is for the adviser to influence the decisionmaking.

[22] Some international groups purportedly treated the outcome as a Farabundo Martí National Liberation Front success, since it was not defeated (Rosello, 1993, p. 107).

[23] Rosello, 1993, p. 104.

[24] Chris Paul, Colin P. Clarke, Beth Grill, and Molly Dunigan, *Paths to Victory: Lessons from Modern Insurgencies*, Santa Monica, Calif.: RAND Corporation, RR-291/1-OSD, 2013, p. 51.

exposure . . . to US military professionalism, respect for human rights, and apolitical attitudes" that had the largest impact on this transition.[25]

This "values transfer," which had been evident to some degree in our drug missions, was a consequence of the fact that our Special Forces and a smaller number of selected advisers from the Marine Corps and other Army branches worked, lived, trained, and operated with our partners.[26] The close relationship was sealed by the sacrifice of a handful of advisers who were killed in combat alongside their El Salvadoran teammates.

The importance of this professionalization should not be understated. Indeed, at the peace talks in 1992, the leadership of the Salvadoran insurgency cited the stationing of U.S. advisers within El Salvadoran brigades as the turning point in the war, as it stopped many of the abuses the military was allegedly inflicting on the people.[27] Further, despite being fiercely anticommunist and a long history of political control, the El Salvadoran military supported the negotiations and enabled the peaceful transition of civilian control, which has been attributed to the deliberate efforts by the United States to shape the composition of the military's leadership.[28]

Professionalization might not have given the El Salvadoran government a decisive victory of the insurgency, but it gave much-needed time for peace to succeed. It created the "stable political and social conditions in which the leading actor, the host country's government, could function productively [and thus] promoted public confidence in the government's ability to govern well."[29]

---

[25] Rosello, 1993, p. 105. Rosello also writes, "Unfortunately, no studies have been conducted to assess this seeming transfer of values, so it is difficult to prove."

[26] This "values transfer" was reportedly also a key mechanism in the later success of U.S. assistance to the Colombian military and Iraqi special operations forces, which similarly relied on U.S. Special Forces—critically, both these partnered forces acted more ethically than units trained by other U.S. military elements (personal communication with Lieutenant General [retired] Ken Tovo).

[27] Rosello, 1993, p. 105.

[28] Hennelly, 1993.

[29] Rosello, 1993, p. 102. The author is also emphatic that "US military and economic assistance did not win the war in El Salvador."

Ultimately, it was the political reform that allowed for the success of the settlement,[30] and the substantial amounts of U.S. assistance ensured that a civilian government would stay in power.[31] Although others—such as the 1984 Kissinger Commission—had recognized this well in advance, the Department of Defense was late in recognizing the "political and economic basis of the civil war" and the importance of a strategy that "explicitly linked countering the insurgency to attaining political legitimacy, which in turn dictated an end to human rights abuses."[32]

Two important points need to be made though that attest to the importance of such seemingly small but effective missions. First, the patience, investment, and modest sacrifice of the United States was appreciated by the El Salvadoran people. Second, the El Salvadoran special operations units would provide capable partners to the United States after the terrorist attacks on September 11, with El Salvador deploying special operations units half a world away as part of the Combined Joint Special Operations Task Force in Iraq for several years.

Strategic Success by Accident: Inadequacy of Existing Models and Structures for This Population-Centric Conflict

Despite our success in El Salvador, it is now clear that the United States did not have the capability to accurately forecast what operations in El Salvador would achieve, despite the then-recent experience in Vietnam. For one, the United States had no idea of the scope that the El Salvador mission would require. As an example, an assessment from the commander of U.S. forces in El Salvador—which concluded that some $300 million in military aid and as long as five years would be

---

[30] See Evans, 1997, for a discussion of the importance of democratization.

[31] Hennelly, 1993, p. 66.

[32] Schwarz, 1991, p. 11.

required[33]—might be considered a bit optimistic in hindsight, as "success" was only achieved after 12 years and a $1 billion investment in military aid (and $5 billion in other types of aid). But "Reagan administration officials" disregarded even this relatively optimistic assessment as "unnecessarily bleak."[34]

We also still lacked the ability to understand the enemy.[35] In 1992, immediately before the peaceful settlement that we would later chalk up to a success, our most-honest assessments still indicated that "the war in El Salvador [was] still a bloody, draining conflict with a clear victory for either side unimaginable."[36] These assessments focused on the fact that the Farabundo Martí National Liberation Front had not "suffered the significant defeats, the large-scale defections, the weakening of their rural support, and the increase in active support for the government and armed forces of El Salvador that would signal the insurgency's decline."[37] Further, contemporaneous observers suggested that a possible settlement with the Farabundo Martí National Liberation Front would be an indication that the United States had failed, at least militarily, and that negotiation was a reflection that the government could not defeat the insurgency.[38]

We similarly did not understand our partners, the bulk of whom had (as of 1988, during my last deployment) neither the will nor the aptitude to be counterinsurgents. Many of their infantry officers, a high percentage of whom had trained in the United States, simply did

---

[33] Fred Woerner, *Report of the El Salvador Military Strategy Assistance Team*, Washington, D.C.: U.S. Department of Defense, 1981, as described in Schwarz, 1991, p. 2.

[34] Schwarz, 1991, p. 2.

[35] The "major flaw" in the analyses before the settlement was "that they failed to take into account the circumstances and nature of the [Farabundo Martí National Liberation Front] guerrillas" (Hennelly, 1993, p. 64).

[36] Schwarz, 1991, p. 3.

[37] Schwarz, 1991, p. 3.

[38] General Maxwell Thurman, then–SOUTHCOM commander, lamented in 1990 that government would not be able to defeat the insurgency, and that negotiation had become the only possible way forward (Michael Gordon, "General Says Salvador Can't Defeat Guerrillas," *New York Times*, February 9, 1990). The assessment that this was a sharp reversal is from Schwarz, 1991, p. 4.

not agree with the U.S. approach for the conflict and wanted a "more rapid, and purely military, conclusion to the war."[39] Some analysts attributed this proclivity to training provided by the United States, which "did not adequately reflect the unique circumstances in El Salvador" and focused on conventional maneuver rather than the Salvadoran insurgency.[40] This was a particularly pernicious challenge for the more educated elements of their military. As an example, the Salvadoran pilots being trained by U.S. intelligence agencies (mostly English speakers with training in the United States) paid only lip service to our efforts to train them on how to distinguish between regular *campesinos* and insurgents.[41] And the more junior people we trained, frequently the "smart kids" from high school, had no understanding of either warfare or what was going on in their country.[42]

It is perhaps unsurprising then that it was difficult for us to get members of the El Salvadoran military to use the counterinsurgency training they received, as they preferred conventional approaches.[43] I do wonder whether the country's military ever fully understood that peace would eventually involve repatriating entire swaths of the country, and the people who inhabited them, which had been lost to the insurgency.

---

[39] Michael Childress, *The Effectiveness of U.S. Training Efforts in Internal Defense and Development: The Cases of El Salvador and Honduras*, Santa Monica, Calif.: RAND Corporation, MR-250-USDP, 1995, p. 28.

[40] Childress, 1995, pp. 32–33. Other analysts reported a similar difficulty with the U.S.-provided training (e.g., Bacevich et al., 1988, pp. 14–15; Michael Sheehan, "Comparative Counterinsurgency Strategies: Guatemala and El Salvador," *Conflict*, Vol. 9, No. 2, 1989, quoted in Childress, 1995, p. 32).

[41] This observation and that in the following sentence are based on personal communication with Andrew Liepman.

[42] These were typically the kids from the upper social classes that had the pull to get their kids into jobs that were perceived as better. I should note that doing well in high school only meant so much in the El Salvadoran military, and many folks who did poorly in high school excelled during professional military education as either an officer or a noncommissioned officer.

[43] Childress, 1995, p. 31.

Most just wanted to kill the insurgents, and they frankly did not much care whether a few farmers were killed during operations.[44]

Finally, it is now widely accepted that our limited military footprint, the 55-person "speed limit" that was an artifact of the then-recent experience of the United States in Vietnam, contributed to our overall success in El Salvador. This was the conclusion of Ambassador Pickering, who concluded: "[O]ne of the things that helped us the most . . . was the limitation we imposed on ourselves, in order to gain congressional confidence in our approach, on the number of U.S. military people we had. *In the last analysis I would judge that that was an ingredient for success rather than failure.*"[45] Many of those who pushed for a larger U.S. presence on the ground during the mission would later conclude that this constraint was critical to the success of U.S. efforts,[46] which has been described in more recent years as "benign neglect."[47]

Our small-footprint approach helped the United States limit the political pressures at home that are normally associated with being at war. Leaders spoke of a "Vietnam syndrome" that would keep the United States out of future population-centric wars or counterinsurgencies. It was the zeitgeist of the times, a brake on the United States sending soldiers into harm's way. The result was that it would be the El Salvadoran soldiers who would do most of the bleeding in this fight. The art of the advisory effort was understanding what was possible,

---

[44] Personal communication with Andrew Liepman.

[45] Manwaring and Prisk, 1988, p. 405 (emphasis added).

[46] Colonel James A. Steele, the U.S. military group commander in El Salvador from 1984 to 1986, made the same observation about the El Salvador experience, concluding: "Nobody has cursed the 55-man limit more than I probably have in the last two and a half years, but I just have to tell you that doing it with a low U.S. profile is the only way to go. If you don't, you immediately get yourself into trouble, because there is a tendency for Americans to want to do things quickly, to do them efficiently—*and the third step in that process is to do it yourself*" (Manwaring and Prisk, 1988, p. 407 [emphasis added]).

[47] A key characteristic of the experience in El Salvador is what has been described as "benign neglect," in that the formal U.S. role was always extremely limited in large part because "it simply did not reflect the American way of war and therefore any serious support would have been a diversion from more pressing military problems." This benign neglect gave special operations the operational maneuver space needed to be successful in the campaign (Rothstein, 2007, p. 279).

as El Salvador's conscript army meant that it must be trained and employed differently from the U.S. Army.

From my perspective, the "speed limit" helped reach a settlement by limiting attempts to accelerate what I believe was the natural pace of the counterinsurgency campaign and our supporting operations. Too big a presence and the supporting nation becomes part or most of the problem. Too big and the American tendency is to grab ownership of the problem. Too big and the host nation naturally steps back and lets the supporting nation bleed for it. The same goes for money—too much is worse than not quite enough.

It is a characteristic of the mind-set of the American way of war that increasing the number of troops and amount of money will shorten the war. In the physics of conventional war, that may be true—indeed, "surges" are part of the math of such warfare. It would be my experience that it is not true in population-centric wars in which the United States supports a nation's counterinsurgency efforts, finds itself the counterinsurgent force, or must conduct unconventional warfare of its own. There, the sciences of anthropology, sociology, and psychology dominate and dictate a different calculus and timeline.

CHAPTER FIVE

Panama and the Transition from Traditional to Irregular

Beginning in the summer of 1989, still a promotable captain though now assigned as the operations officer for 7th Special Forces' Panama-based 3rd Battalion, I was tasked with leading our battalion's planning for a possible U.S. invasion of Panama.[1] At the time, our battalion was unquestionably the most capable foreign internal defense unit in the Department of Defense and arguably its most fertile laboratory for building irregular warfare expertise. We also just so happened to be located in Panama, where we had been based for almost three decades, and were thus the ideal candidate to conduct unconventional warfare in support of Operation Just Cause.

Our battalion would form the backbone for Task Force Black, which would be the only irregular warfare component in operations in Panama.[2] Task Force Black would leverage its placement, access, and knowledge of the country and its people to enable special reconnaissance and direct-action missions in support of the initial invasion. In the aftermath of the removal of Panamanian President Manuel Noriega, this task force would then become the primary effort in securing the

---

[1] This last assignment, which culminated my 3.5-year assignment with 3rd Battalion, 7th Special Forces, was as an operations officer (S-3).

[2] This task force would combine our battalion's staff with that of SOCSOUTH, and our two peerless foreign internal defense–oriented companies and language-qualified counter-terrorism company would be joined by the 617th Special Operations Aviation detachment to form a formidable, broadly expert irregular warfare force.

peaceful surrender of Panamanian security forces as part of Operation Promote Liberty.

Operation Just Cause deposed President Noriega, a narcotrafficker who had seized power in 1981.[3] Although he was previously an ally of the United States with close ties to U.S. intelligence and law enforcement agencies,[4] U.S. support for Noriega began to deteriorate by the mid-1980s. In November 1987, in the wake of Noriega's suppression of pro-democracy riots and an attack against the U.S. Embassy in Panama City, the United States terminated all economic and military assistance to Panama. The following March, as Noriega was increasingly soliciting support from Soviet-aligned Cuba, Libya, and Nicaragua, contingency planning for Noriega's removal began.[5] President George Bush would give the order for the invasion in December 1989, in the wake of months of provocation by Noriega's security forces targeting U.S. citizens in Panama that culminated in the death of a U.S. Marine.

Operation Just Cause was, by any measure, a tremendous success. Noriega's security forces lost control of Panama City by sunrise of the first day, and all but his most loyal paramilitary forces surrendered within a week of the initial invasion. By January 10, even these forces had surrendered, Noriega was in U.S. custody, and the United States had installed a democratically elected president.[6]

Although successful, this success had less to do with U.S. planning for transition and much more to do with our country's shared history and symbiotic relationship with the Panamanians. The initial phase was well rehearsed and well executed, demonstrating the potency

---

[3] His predecessor died suddenly in a plane crash that many suspected was an assassination.

[4] U.S. Senate, *Drugs, Law Enforcement and Foreign Policy: Hearings Before the Subcommittee on Terrorism, Narcotics, and International Communications and International Economic Policy, Trade, Oceans, and Environment of the Committee on Foreign Relations, United States Senate,* Washington, D.C.: Government Printing Office, 1988.

[5] Ronald H. Cole, *Operation Just Cause: Panama,* Washington, D.C.: Joint History Office, 1995, p. 7.

[6] President Guillermo Endara had received the majority of the vote in a May 1989 election, but Noriega annulled the results of the election.

of violence, speed, and overwhelming firepower in winning the tactical fight. What we lacked was a cogent plan for exploiting tactical success, a plan that would map out how this unquestionable tactical prowess could be transitioned into strategic success. An inability to forecast how a population-centric conflict could emerge in the wake of U.S. tactical successes, as well as to plan and prepare for resistance (whether nonviolent or violent), would be a recurring theme in future U.S. campaigns. There would not be a *transition plan* (as it would later be called) worth the paper it was written on.[7]

Operations in Panama also demonstrated (at least to me) that the U.S. military did not have the concepts necessary for success in population-centric warfare.[8] As an example, we learned the hard way that speed and surprise could be counterproductive under the wrong circumstances, despite being successful during the initial invasion in Panama (and in regular combat operations). I saw this firsthand shortly after the initial invasion, during the very early days of Operation Promote Liberty, when our counterterrorism company and U.S. Army Rangers inadvertently created unnecessary anxiety and uncertainty among the population by employing the textbook approach (e.g., at night, with overwhelming firepower) in securing the surrender of what was a neutral, if not friendly, city. This created a real risk that the population could turn hostile. If we continued to use such tactics across the country, we would have threatened the overall long-term success of the mission. The company and battalion leadership recognized immediately that we had to find a better way.

---

[7] The stalemates in Afghanistan and Iraq years later were, in my opinion, at least in part a consequence of this lack of planning and, at a more fundamental level, a lack of understanding. The U.S. Army would later conclude: "Successful Army units use a projected transition plan and timeline, developed with host-nation participation, to redress the [host nation's] concerns about U.S. occupation, secure host-nation buy-in, and mitigate any tendencies to become overly dependent upon the United States" (Army Techniques Publication 3-07.5, *Stability Techniques*, Washington, D.C.: Headquarters, U.S. Department of the Army, August 2012, p. 5-2).

[8] It is the concept that drives development of a service's doctrine, organization, and training needed to employ the concept.

Planning for Regular and Irregular War: My First Special Operations Campaign Plan

My Panama-based Special Forces battalion went into serious preparation for supporting an invasion in the summer of 1989. Our battalion was the remnant of a continuous and, at times, robust Special Forces permanently assigned presence in Panama since the 1960s.[9] Amid rising tensions between Noriega and the United States, we were tasked (along with others) to use our placement and access to get inside Noriega's decision cycle, to put some uncertainty into his planning, and to be prepared to support an invasion.[10] We remained engaged in our ongoing missions in Central and South America, mainly to avoid passing our assigned tasks to other units for a contingency that had a high likelihood of not happening. The battalion also had to maintain its operational tempo to avoid tipping off the Panamanians.[11]

Our battalion's planning directive, which came indirectly from the then-new Joint Special Operations Command, had two components. During the initial phase of the invasion, our battalion was tasked to be prepared to conduct special reconnaissance missions on more than 30 targets across the country. After the initial invasion, we would then be supported by the group headquarters and an additional battalion from 7th Special Forces, both arriving from Fort Bragg, in securing the surrender of Noriega's security forces.[12] This second phase of the U.S. mission was the lesser-known Operation Promote Liberty,

---

[9]   In 1972, 3rd Battalion, 7th Special Forces, was established when the Panama-based 8th Special Forces Group (Airborne) was deactivated. 8th Special Forces first arrived in Panama in 1962.

[10]   This responsibility was in addition to our routine deployments to El Salvador, to Bolivia and Peru for the drug war, and throughout Latin America for training missions, which we continued while planning for what would eventually be Operation Just Cause.

[11]   In retrospect, it is a credit to the battalion and company commanders who were remarkable in managing missions and personnel, as the battalion was ready to do its part when the call for Operation Just Cause finally came.

[12]   The Task Force Black plan for Operation Promote Liberty called for the 7th Special Forces Group (Airborne) headquarters and its 2nd Battalion to deploy to join our battalion. They would arrive shortly after the initial invasion and stay in theater for just over a month, then transition the mission back to our battalion.

the stabilization mission that followed the initial invasion, in which 7th Special Forces would arguably become the main effort.

As the operations officer for the battalion, it was my task to coordinate our battalion's planning. Our first challenge was planning for the initial phase of operations. Two of our three companies, which included a total of 12 Special Forces teams, would be asked to simultaneously execute the more than 30 strategic reconnaissance missions called for in the plan. Although our battalion was arguably the most experienced and proficient foreign internal defense and combat advisory unit in the U.S. military, our teams were a bit out of practice when it came to strategic reconnaissance. The difficulty was compounded by the fact that few of our soldiers had the Top Secret clearance necessary to know the details of missions other than the one they had been assigned. In the end, we divided into teams of three to six Special Forces operators and then compartmented them from each other during a multiday planning event.[13] Our counterterrorism company, which already had the requisite assault, sniper, and reconnaissance teams for its assigned role in the operation, would face the more modest challenge of balancing ongoing requirements for training of U.S. partners in Latin America against readiness for potential direct-action missions in Panama.

We had briefed our plan that summer to our colleagues at the Joint Special Operations Command, my first exposure to the fairly new U.S. national counterterrorism capability. We would end up spending months planning together, and the mission would have elements from across the special operations community, playing roles befitting their strengths. The Rangers, SEALs, and special mission units would seize airfields and rescue hostages, while we (the Special Forces contingent) would focus on the strategic reconnaissance missions across the country and then, in a few instances with conventional units, secure the peaceful surrender of Noriega's troops.

We also spent months rehearsing together.[14] The Army's special mission unit frequently rotated its in-country elements, and we became

---

[13] Most of the planned strategic reconnaissance missions were never executed.

[14] R. Cody Phillips, *Operation Just Cause: The Incursion into Panama*, Washington, D.C.: U.S. Army Center for Military History, 2004.

very familiar with the operational piece as we rehearsed with each new element when it arrived, marshaling sometimes three times a week in the run-up to the invasion. These rehearsals were a requirement of General Colin Powell, then–Chairman of the Joint Chiefs of Staff, who stressed that "plenty of manpower and rehearsals" were necessary if we hoped to take down Noriega's forces with limited casualties.[15]

Rehearsals like this were understood to be necessary by everyone in the unit but were unusual for us. Most of our real-world deployments and down-range advisory and training missions did not lend themselves to being rehearsed.[16] The value of the coordination that inherently comes with repeated rehearsals was obvious. The memory of the failed Operation Eagle Claw was still vivid in everyone's mind—reportedly including President Bush's, who wanted "assurance that it would not backfire as had the attempted rescue of U.S. hostages in Iran during the Carter administration."[17]

In the run-up to the invasion, our battalion had planned to be the headquarters and staff for Task Force Black. But the Special Operations Command South (SOCSOUTH) commander decided to combine our battalion with his small but rank-heavy staff to form Task Force Black instead.[18] I was to be the ground operations offi-

---

[15] Cole, 1995, p. 18. In the end, none of the contingencies that we rehearsed was ever executed, and I would later wonder whether our time might have been better spent exploring irregular war options that could have either supported the invasion or provided an alternative approach for dealing with Noriega.

[16] To be sure, we did rehearse immediate action drills, linkups, and other mechanics, as well as the very few direct-action missions the unit was handed.

[17] Cole, 1995, p. 29. In retrospect, these rehearsals were particularly important for the U.S. special operations community because cooperation between the two halves of our community had been limited during the previous decade. A fissure had developed between the "hyperconventional" and irregular wings of U.S. special operations in the decade after Operation Eagle Claw, as a result of a growing disparity in resourcing and the allocation of favored missions to one force over the other.

[18] When we were reinforced by the Fort Bragg–based 7th Special Forces' battalion, the group headquarters and its commander deployed along with them. Our group commander, also a colonel, would himself work for the SOCSOUTH commander. The overall invasion was commanded by Lieutenant General Carl Steiner, the XVIII Airborne Corps commander who was both Special Forces–qualified and had time within the Rangers.

cer for Task Force Black, but I would technically be working for the SOCSOUTH commander's operations officer rather than my own battalion commander. This novel joint headquarters maximized our peacetime chain of command while augmenting SOCSOUTH's very short-staffed headquarters. The result was direct access to their organic aviation capability and a senior leader with enough rank to be heard within the Joint Special Operations Command headquarters.[19]

At the time, I resented the arrangement, as it was our battalion's plan and I really thought we did not need SOCSOUTH's help. Further, I had supreme confidence in my battalion commander, Lieutenant Colonel Roy Trumble, and did not look forward to taking orders from any of the staff officers at SOCSOUTH. I concede now that the consolidation made sense. I was naive in how controversial and contentious these chain of command arrangements can be and did not appreciate the value of SOCSOUTH's access and rapport with the four-star U.S. Southern Command (SOUTHCOM). I was also too prideful and worried that our battalion's good work would be expropriated.[20] Many years later, I would use a very similar arrangement for operations in Iraq when I commanded Task Force Viking while the commander of 10th Special Forces and again with Operation Willing Spirit in Colombia when it was my turn to be the SOCSOUTH commander.

In the wake of the initial invasion, in which Task Force Black executed only a few of our initially planned (H-hour) missions, we transitioned to focus on securing the peaceful surrender of Noriega's security forces. As events played out, 7th Special Forces became arguably the main effort following the initial invasion, playing a critical role

---

[19] Colonel Jake Jacobelly, the SOCSOUTH commander, also brought with him elements of 4th Psychological Operations Group and a variety of air assets (Douglas I. Smith, *Army Aviation in Operation Just Cause*, Carlisle, Pa.: U.S. Army War College, April 15, 1992). Because Task Force Black was subordinate to the two-star Joint Special Operations Command leadership (Major General Wayne Downing), having a full colonel leading Task Force Black did have its advantages. However, there is no doubt in my mind that Lieutenant Colonel Trumble, our battalion commander, would have done fine representing Task Force Black as its commander, as we had initially planned.

[20] It was at about this point that it was dawning on me that the Army is a team sport and, as in most jobs and professions, much can be done without worrying about credit.

in the stabilization of Panama City and "became the sole U.S. presence with responsibility for almost anything that was done in the name of the new government."[21]

This stabilization mission, which was the core of Operation Promote Liberty, was nonstandard but clearly fit under the broad definition of *irregular warfare*. Task Force Black's approach relied on a combination of lethality, understanding, and empathy to subdue Noriega's security forces, an approach that our Special Forces were uniquely well qualified to execute. In Panama City, where the United States maintained a large conventional presence, a handful of our Special Forces teams quickly moved into the most-dangerous neighborhoods, canvasing them to determine sentiment and gauge the threat.[22] Outside Panama City, our teams would establish a presence in a series of preidentified cities and towns, assessing whether that municipality contained hardcore Noriega supporters and quickly rehabilitating remnants of the Panamanian military and police judged to be at least benign toward the new Panamanian government of President Endara.[23]

Speed, Surprise, Overwhelming Firepower . . . and Good Preparation: Success in the "Traditional" Warfare Component of Just Cause

When the order for the operation finally came, the preparation, movement, and quartering tasks were identical to the rehearsals, and we were well prepared. We simply boarded the helicopters, as practiced, when they showed up at our battalion headquarters (at Fort Davis on

---

[21] John T. Fishel, *The Fog of Peace: Planning and Executing the Restoration of Panama*, Carlisle, Pa.: Strategic Studies Institute, U.S. Army War College, April 15, 1992, p. 47.

[22] We were particularly concerned about Noriega's so-called dignity battalion, which was a group of paramilitary forces that were fiercely loyal to Noriega and were seen as one of the primary obstacles to U.S. success in Panama (Sam Fulwood III, "Combat in Panama: Dignity Battalion Still Lurks in City Shadows," *Los Angeles Times*, December 22, 1989).

[23] This mission was the responsibility of teams in 3rd Battalion's Alpha and Bravo Companies and those from 2nd Battalion. In some locations, these forces would be eventually used to relieve other U.S. forces.

the Atlantic side of the Panama Canal) and 30 minutes later landed at our prepared tactical operations center in a hangar at Albrook Air Force Base (on the Pacific end of the canal), which would serve as the primary command post for Task Force Black's portion of the invasion. All we had to do was update our maps, to ensure that we had the latest intelligence available, and we were ready to go. For most involved, it just appeared to be another rehearsal.

Almost immediately, several of the reconnaissance missions assigned to our task force became direct-action missions. When one of our reconnaissance teams learned of a brigade of Noriega's heading toward the city, which would endanger the 82nd Airborne's imminent assault on the international airport and allow Noriega to reinforce his defense of Panama City, our battalion quickly assembled a strike force to interdict the brigade at the Pacora River Bridge.[24] A firefight began almost immediately on infiltration, and our Special Forces soldiers would hold the bridge the entire night.[25]

A similar operation unfolded in the mountains just to the north of Panama City (in Cerro Azul) when one of our teams was instructed to disable and secure a local TV station. The mission was to disable the transmitter in such a way as to make its return to operation quick and easy. The team secured the services of a Panamanian TV technician, who received a quick lesson in how to fast-rope from a helicopter and then infiltrated with the team. After securing a key electronics component, the team was extracted with no issues.

Our counterterrorism company also proved its worth during the early hours of the operation.[26] Responding to a last-minute tasking,

---

[24] Major Kevin Higgins quickly and quietly assembled an assault force from multiple Special Forces teams in his company to conduct a platoon-sized raid, a very nonstandard approach for Special Forces, to secure what proved to be a strategic choke point. Its success was a testament to his leadership and the professionalism and adaptability of the troops executing the mission. They had seemingly disproved the notion that, while three Ranger squads on mission make a platoon, three Special Forces teams on the same objective is a bar fight.

[25] Phillips, 2004, p. 20.

[26] This concept, which had been developed by Colonel Chuck Fry, a previous SOCSOUTH commander, was then referred to as the Commander's In-Extremis Force and worked directly for the four-star geographic combatant commander. It would be later renamed the

and thus conducting an operation for which we had not developed a plan, this company successfully interdicted a clandestine regime radio station. The objective was in a high rise in downtown Panama City, and the company executed a classic raid: fast-roping onto the roof of the 17-story building, positioning a security team on the street-level entrances, and then disabling the target with thermite grenades before safely disappearing into the night.[27] These companies are now a critical capability in every region, serving as a bridge or hybrid capability between U.S. direct-action and irregular warfare capabilities, and I would see them perform successfully as a force multiplier alongside foreign special operations units both with and without the national special mission units in every theater before I retired.

This small handful of missions—several of which were a variation on the 30 or so that the battalion had planned and a few others were last-minute requirements—was important, but each was probably not essential on its own (at least in my view). The capture of the Pacora River Bridge was the most noteworthy, as the U.S. airborne units that jumped into the airport (then named the Omar Torrijos International Airport and soon after renamed the Tocumen International Airport) would have had a very different story to tell had that Panamanian brigade made it across the bridge. But even then, there was little doubt in the invasion's outcome. The question was how quickly objectives could be taken and loss of life and destruction of property minimized— as it turned out, very quickly and with remarkably limited loss and destruction. In the early morning, five hours into the operation, while watching the downtown Panama City burn, it became clear that the "dog had caught the car." The fairly straightforward part complete, the question was what would happen next.

Crisis Response Elements. This company had trained commando elements throughout the region that would be first responders in their home countries.

[27] One security guard was purposefully wounded, after he decided he'd unholster his pistol as the Americans were leaving. He had been warned in Spanish by an operator that if he kept his weapon holstered, he would not be shot.

Postinvasion Unconventional Warfare: The Peaceful Surrender of Panamanian Security Forces

As initial combat operations came to a close, and having already been reinforced by a Special Forces company from the Fort Bragg–based 7th Special Forces prior to the invasion, we were reinforced by battalion and turned to the business of securing the peaceful surrender of Noriega's security forces. As the ground operations officer for Task Force Black's role in the invasion, I was stuck at Albrook Air Force Base during combat operations. But in the early morning hours of Christmas Day in 1989, the sixth day of the operation, I rejoined my old company, fresh off of its capture of the Pacora River Bridge, for what was essentially the first days of what should be called the *irregular warfare phase* of the special operations mission. In just under a week following the invasion and the successful assault on the bridge, this company would secure the surrender of the Panamanian garrison in the city of David and capture its commander, Lieutenant Colonel Luis del Cid, one of Noriega's closest aides.

We had learned quickly, though unfortunately the hard way, that an "aggressive" approach could be counterproductive. Our counterterrorism company took the lead in securing the surrender of the first major city, a town called Penonomé, as part of this second phase of our mission. It was executed in a prudent manner and in line with doctrinal urban operations tactics, leveraging darkness ("we own the night"), speed, and surprise.

But it created a huge amount of anxiety in the townspeople and uncertainty among those whom we wanted to surrender. A Panamanian soldier was reportedly killed in the raid, which risked creating animosity in a community that had no love of Noriega or his rule. It was something that we could ill afford to repeat. It was a testament to the skills and awareness of the battalion and company commanders and the creativity of the Special Forces senior noncommissioned officers who recognized this challenge immediately and updated the approach for securing follow-on surrenders.

The United States did not want to be occupiers and getting the Panamanians to work with and for us was critical if we wanted to get out of becoming the local jailer, jury, judge, and executioner. So, after Penonomé, we started doing our surrender operations during the day and we literally phoned ahead, which came to be called the "Ma Bell" approach.[28] We had worked with the Panamanians quite a bit, and while the idea of calling them on the phone was unorthodox at the time, it is a no-brainer in retrospect. They were very likely to follow instructions, particularly when we had gunships circling over their compounds, as they had by then seen the pictures of what happened to Panama City. The reality was that most were not particularly fond of Noriega to begin with; most people did not believe that he and his corrupt administration were worth dying for.

Years later, when I led the northern invasion of Iraq in 2003, I anticipated that we would use a similar approach with the Iraqi Army. Building from this success in Panama, our concept for northern Iraq was to turn security over to the Iraqi Army, and all of our intelligence indicated that a reconstituted Iraqi Army and police forces could be coerced and advised into returning to their security and policing roles rather quickly. But within a couple of months of the invasion, the United States would make the fateful decision to leave the old Iraqi Army and police forces jobless around the country. We thus became the occupiers in Iraq, a label we managed to largely avoid in Panama.

On Christmas Day in 1989, I got a "kitchen pass" of sorts from the battalion commander to leave the operations center and join up with Kevin Higgins, who was commanding my old company in the liberation of David. I had deployed to David in 1987 on the last joint U.S. and Panamanian Special Forces exercise (Kindle Liberty) and knew the city well. I offered to assist where I could, given my firsthand knowledge.

It started as a classic special operation. Departing in the dark, early on Christmas morning, we conducted a midair refueling off the coast of Panama with the glow of first light allowing witness to the aerial ballet of black-painted special operations helicopters taking turns

---

[28] The first "Ma Bell" operation was led by Major Gil Perez and a collaboration of his company (A-1/7) and 3-27th Infantry (David S. Hutchinson, *The 3d Battalion 27th Infantry in Operation Just Cause*, Carlisle, Pa.: U.S. Army War College, 1992).

getting gas. The first stop was a university extension campus some 20 miles or so outside town. Higgins made the phone call from the school administrative building that instructed the Panamanian garrison on surrender terms. We then remounted the helicopters, flying over dozens of farms and small towns whose occupants came out waving American flags and landed in the heart of Panama's second largest city, David. I wanted to believe that most waving flags were thankful for what we had done in deposing Noriega, and their cheers fed my belief that they knew we were indeed the "good guys." It could be they simply did not want to be bombed.

The entire garrison was waiting for us in formation when our company arrived, as Higgins had instructed, and the gunship circling overhead ensured that there would be no issues. But the plan also called for the antiaircraft gun at the city airport (an old "quad forty" Bofors) to be pointed down or it would be destroyed by our gunship. The mission was going very well so far, with no shots fired and great cooperation from the Panamanians. Therefore, I recommended to Higgins that I go to the airfield and make sure that the Panamanian security forces there were also complying with instructions. I asked the assembled Panamanian captives whether anyone had a car. After selecting a police officer from among the raised hands, two of us jumped into his nearby car and we were at the airfield 15 minutes later.

As luck would have it, a Panamanian special operator I had worked with during the Kindle Liberty exercise back in 1987 was standing in formation next to the depressed antiaircraft gun. The young lieutenant was easy to recognize: tall for a Panamanian, he was also wearing the distinctive headgear of Panama's Israeli-trained special operations unit. Fortunately for me, he was quick to remember our association, telling me proudly that he had become the head of special operations for that entire region of the border. This admission made him a person of interest and got him a trip to Panama City for questioning, but he was quickly returned and continued to serve his country's new leadership.

After about as much pleasantry as the awkward situation allowed, I asked him where his soldiers were. He stated that they were awaiting instructions and that their actions would be determined by those of the Americans. I explained that we were not there to occupy but simply to remove Noriega and his crew, and we needed his guys to come in and help be a part of what comes next. He agreed and said that some of them were nearby if we wanted them to come in. It was worth the risk in my view, and, on his signal, ten guys drove up in a van, piled out armed to the teeth, and surrendered to us. There was no way we could carry all their weapons, so they remained armed. Looking back, there were a lot of ways to have handled this situation, many perhaps better than what I chose to do, but, regardless, it worked.

The complexities of the mission began to quickly mount. One stuck in my memory is a young Panamanian woman who came up to us very soon after we had landed in town, covered in blood and sobbing, while we were registering our Panamanian "captives." She told us that her brother-in-law had stabbed her sister. The last thing we needed was to become homicide investigators, so we identified a Panamanian detective standing in line to be registered, gave him one of the weapons we had just impounded, and instructed him to go with the lady and do what he ordinarily would.

Our battalion would keep Special Forces in David for several months, but my kitchen pass was up, and I flew back to Panama City that same day with then–Major General Wayne Downing on the MC-130 carrying del Cid. We handed off del Cid to DEA officials at Howard Air Force Base, where they hustled him to another MC-130 that was waiting on the ramp, blades turning, to get him back to the United States.[29] He would be a key witness for the prosecution in Noriega's conviction for drug-related offenses.[30]

Although many books had been written by that time on irregular warfare, few prepared us for the complexity of the mission that was unfolding before us. Major General Bernard Loeffke, a Special Forces–qualified and Vietnam War–era icon, had told me three years earlier in Bolivia that, in Special Forces, "You are paid for your judgment." I

---

[29] For a discussion of the handoff from the Air Force perspective, see Jose F. Jackson, "'A Just Cause,'" *Citizen Airman*, March 1990.

[30] Mike Clary, "Key Witness Against Noriega Sentenced to Time Served," *Los Angeles Times*, July 10, 1992.

saw this at David, and I would see it in the months that followed as we shifted our efforts to winning the peace.

Securing the Peace: Transition into an Irregular Warfare Mission

In the early hours of the invasion, as I watched our gunships just hammering Panama City in the predawn darkness, I realized that we were going to own Panama's problems in very short order. We had planned for the postinvasion transition, in which our Special Forces battalion under Task Force Black would play a prominent role. But as it was becoming clear that our victory was assured, I began to realize how insufficient and incomplete our plans and those of the headquarters above us were. Those streets that we were lighting up with our gunships' tracers were the streets I knew well, having gone to high school in Panama, and our plans did not come close to addressing the complex problem set that was now ours.

The U.S. success in the aftermath of the well-scripted invasion should be attributed largely, in my view, to the professionalism, ingenuity, and grit of the Special Forces teams who would be sprinkled about the country to keep the peace.[31] With the end of combat operations, Task Force Black's primary mission shifted to establishing a presence throughout the country, and our battalion returned to its garrison on the Atlantic side of Panama as we reconfigured ourselves for our role in Operation Promote Liberty. This new mission required us to conduct continuous assessments on potential hostile activity, mitigate local problems so that they would not become significant enough to require a larger U.S. response, and provide security to the local population when necessary. For this last local security task, we would leverage rehabilitated members of Noriega's security forces whenever possible.

There would be some significant ongoing learning for us, but we developed a standardized approach relatively quickly. Our teams would

---

[31] Conventional units were sent to some of the towns, but Special Forces teams were the most significant presence in the countryside.

enter a town, conduct reconnaissance and identify local security forces, and then secure the surrender of the garrison. It became apparent that knowing the local mayor, business leaders, and other local influencers were key to our success. So, each team was assigned an area of operation, and it was responsible for getting as complete an understanding of the human terrain in the key towns and cities in that area as possible. Within five years, we would use a similar approach to great effect in both Haiti and Bosnia.

Significant parts of the country would not get a dedicated Special Forces team; for those, we instead deployed an ad hoc, nonstandard expeditionary capability. This small and tailored but expanded unit was built around one of the Special Forces teams from our counterterrorism company,[32] which we augmented with military intelligence and civil affairs soldiers. These military intelligence–civil affairs combined teams, which we called MICA teams, would deploy to larger isolated towns to conduct stability assessments, providing us a mechanism to understand what was transpiring outside the areas immediately affected by the invasion. The intelligence soldiers would do interviews to identify potential bad actors among the security force and population, and any reportedly abusive or corrupt Panamanian we could find would be returned to Panama City for further investigation. At the same time, civil affairs soldiers would assess local conditions looking for urgent needs that, if not met, could lead to security issues.

However, despite our successes, the difficulty of administering the peace so soon after conflict would be apparent. I experienced this firsthand in the tiny seaside village of Puerto Obaldía, nestled just a few kilometers from the Colombian border on the Atlantic coast. Our augmentation from Fort Bragg had not yet arrived, so the whole of Panama, less the fairly small conventional footprint, remained ours to assess and secure. While maintaining my primary duty as the battalion's operations officer, I would for a short time command an unusual Special Forces company that we formed as part of our plan for administering the peace. This company was an Operational Detachment B, the wartime term for a deployed Special Forces company headquarters.

---

[32] This counterterrorism company was 3rd Battalion's "Charlie" Company.

But what was unusual was that I would "borrow" a Special Forces team from each of the battalion's three companies, including the counterterrorism company, so that our battalion would have four functional companies, with one company assigned to each of Panama's four quadrants. The newly formed company, Operational Detachment B 777, was assigned the northeast quadrant of the country, where we would be responsible for maintaining situational awareness and, as best we could, administering the peace.[33] We planned to do this through recurring trips to key communities in our quadrant, most of which could only be reached by air or sea.

We knew that Puerto Obaldía was a major smuggling nexus for the Colombians and a place where the drug dealers, gold smugglers, and even some members of Colombia's Fuerzas Armadas Revolucionarias de Colombia (FARC) would come for a little bit of rest and recreation. It had a small garrison, all of whom came from the village. They were beat cops, knew the deal under Noriega, and turned a blind eye on illicit activity that was endemic to the region.[34] Very shortly after we helicoptered in with two of my new Special Forces teams for the initial assessment, the garrison readily swore allegiance to the new government.

But, perhaps emboldened by our presence, this garrison also committed to ending the illicit behavior in their community. This would prove fatal for the senior Panamanian police sergeant I left in charge. Judging the security situation as nonthreatening to the United States, I elected not to leave a team behind. I learned a month later, when one of my teams returned to check the town, that this sergeant had been killed when he tried to confront some of the criminals smuggling goods in and out of Colombia. We do not know whether he had tried

---

[33] The four quadrants had been formed by bifurcating the country by the Panama Canal and then assigning one company to the high ground and another to the jungle on each side of the country. Operational Detachment B 777 would be responsible for the San Blas Islands to the Colombian border on the Atlantic side and could be commanded from our base at Fort Davis, which allowed me to command the company while maintaining my role as the battalion's operations officer.

[34] We found warehouses with champagne and other luxury goods, dispelling any doubts about who might be coming through.

to radio back for help, but what was evident was that our intervention in local affairs had disrupted the regular order of things. I had left thinking the security situation was handled, not knowing that we had put this small group of Panamanian patriots in mortal danger. I should have left a Special Forces team there.

Senior military leaders turned to Special Forces to support this stabilization mission in part because they did not have enough conventional forces to secure the entire country in a traditional fashion. But our Special Forces teams proved especially adept at getting both police and local government functioning, in towns from Bocas del Toro to the Darién Gap jungle. Their hallmark was the exercising of good judgment in dealing with an endless list of unplanned scenarios—judgment for which Special Forces, civil affairs, and psychological operations soldiers, and later the marines of U.S. Marine Corps' Special Operations Command and others, are specifically assessed. These were ill-defined "presence missions," but the reality was that they were irregular warfare operations and activities. They were anathema to most conventional commanders but, as the United States would learn in the irregular conflicts to come in Haiti, Bosnia, and Kosovo, all too necessary.

Winning the Peace: More Luck Than Planning in the U.S. Strategic Success

Operation Just Cause successfully removed Noriega from power and neutralized his security apparatus, allowing the United States to install Endara, who had handily won the presidential elections earlier that year, as the president of Panama. It demonstrated the potency of U.S. military power and the effects that the joint force and its new special operations component could generate.

What was unusual about Operation Just Cause, as compared with what would be the U.S. experience in Afghanistan and Iraq, was that it was not followed by an insurgency. This was, in my view, because Noriega and his regime were so reviled, the cultural ties between Panamanians and Americans were fairly strong,[35] and the United States wisely left much of the bureaucracy and security forces in place. It also helped a lot that we had given up the canal at that point,[36] and Panamanians realized that their economic well-being was tied to America and a peaceful Western Hemisphere.

Although some may debate the reasons for our success in Panama, it is clear that our success was not because we were well prepared for how to secure the peace. At the time, our planning for the operation was singularly focused on winning the initial fight, and we did not have a clear strategic-level plan for what to do after the initial phase of the invasion. This was even true for the irregular warfare-focused Task Force Black, as our emphasis had been on the direct-action missions assigned to us for the initial phase of the invasion. We knew that our knowledge of Panama, language skills, and ability to operate in small, self-sustaining, and lethal teams would play roles in the aftermath. But we did not have a transition worth speaking of.

In Panama, the United States made the decision to leave much of the existing governance and security structures in place, which turned out to be the right decision. The signs of what could have happened if U.S. policymakers and military leaders had made different decisions were all too apparent. We saw this in the anger among peasant farm-

---

[35] The United States established a persistent military presence in Panama beginning in 1903. Facing a Colombian government unwilling to lease the land needed for the construction of the Panama Canal, the United States supported a successful Panamanian bid for independence from Colombia and then signed a 100-year lease with newly forming Panama. The United States would maintain control of the Canal Zone, a five-mile buffer on either side of the canal, from the completion of the canal in 1914 until its transfer to the Panama Canal Authority in 1999.

[36] U.S. control of the Canal Zone became a trigger point for domestic instability in Panama beginning in the 1960s. In 1964, in the wake of rioting at the canal that left more than 20 Panamanians dead, Panama broke off diplomatic relations with the United States until the treaty governing the canal was renegotiated. Initial negotiations failed, and by 1968 the newly elected President Arnulfo Arias was demanding the immediate return of the canal to Panama. Finally, in 1977, the United States and Panama began a process that would eventually cede full control of the canal to Panama by 2000. This treaty was signed on the Panamanian side by Omar Torrijos, who had become the de facto Panamanian dictator in 1968 when he seized power from Arias in a coup d'état, and a key component of this treaty was that Panama begin pro-democracy reforms.

ers after U.S. forces confiscated their machetes (which were viewed as potential weapons) and among the Panamanian military professionals who were watching for American mistreatment as an excuse to incite violence against Americans. If we had elected to disband and then completely rebuild Panamanian security forces and replace much of the existing governance structures, as we would do in Iraq just over a decade later, the situation likely would have turned out very different.

The reality is that domestic civilian instruments of governance are typically badly damaged by military operations. Restoring or replacing the local, organic structures is very difficult and is made more so the stranger and more foreign the culture. Instability quickly follows if the basic needs of the citizenry are not restored, and the liberator quickly becomes the occupier in the eyes of the population. Although the obvious answer is transition as quickly as you can to an indigenous structure, this has proved extremely difficult. This is made even more challenging when the invader wants to make substantial changes in how the now occupied society is governed (as compared with pre-war). Our proclivity to mirror image, impose our own values, or pursue social justice causes can make it harder still.

Variations on the approach that we used in Panama—sending Special Forces teams into uncertain situations to mitigate, report, and fight (if necessary)—would become the hallmark of nearly every U.S. ground war, conflict, or peace enforcement operation that followed. Like in Panama, there was typically no coherent plan for what followed the invasion and occupation. As a result, instead of a well-rehearsed orchestration of military and civilian agencies working to secure the peace, a whole-of-government "symphony," what followed would invariably be a "jazz jam" session. It is reasonable to think that the variables are too many and uncertainty too high to really plan for the aftermath, which would explain why such units as our Special Forces teams—built to wade into uncertainty, mitigate problems locally, and report coherently and in depth—would be in such high demand. Special Forces teams do jazz.

Looking back, our greatest problem was the lack of a conceptual framework for making the transition from a traditional war (the initial invasion) to the inevitable irregular war (securing the peace).

Usually, there are ways to capitalize on indigenous capacity, to leverage the goodwill toward Americans that has become almost a global phenomenon and to call on near universally held human desires. But such concepts were not, and are not, in the books. We simply did what we thought best, which proved to be good enough because we knew Panama very well. We would not have this advantage in conflicts to come, and our lack of adequate preparation would prove a persistent challenge in winning the peace in the wake of U.S. tactical successes.

CHAPTER SIX

The Decade of Delusion and My Pentagon Wars

I spent the First Gulf War at the Command and General Staff College. Many of my classmates had yet to see combat, and they watched with disappointment and in some cases despair as their former units marched off, without them, to the first significant conventional campaign since Korea. I had no grounds for complaining, as I had just completed a storybook tour with 7th Special Forces, with deployments in Argentina, Bolivia, Chile, El Salvador, and Venezuela and an instructive role in Operation Just Cause. But my own frustration would come soon enough, as America's decisive success against Saddam Hussein's forces in this new war would have a profound effect on America's irregular warfare capability.

The First Gulf War heralded the emergence of American supremacy in conventional warfighting, demonstrated to the world by the quick work made of the Iraqi Army in Kuwait in 1991. The dramatic success of America's new stealth and increasingly reliable precision-strike capability validated the belief that "superior technology and tactics . . . would dominate modern warfare."[1] The war demonstrated, in sharp contrast to the Vietnam experience, that the United States could now "win quickly, decisively, with overwhelming advantage, and with few casualties."[2] Further, the end of the Cold War marked the arrival

---

[1]   Anthony Cordesman, "The Real Revolution in Military Affairs," Center for Strategic and International Studies, August 5, 2014.

[2]   Attributed to General John Michael Loh during an Air Force Association symposium held on January 31, 1992 (Air Force Association, *Strategy, Requirements, and Forces: The Rising Imperative of Air and Space Power*, Arlington, Va., 2003, p. 19).

of what was perceived (at least in the United States) as a new and hopefully more peaceful world.[3]

Our success in Operation Desert Storm, coming in the wake of the successful conclusion of the Cold War,[4] birthed President George H. W. Bush's "new world order." This new order would purportedly bring with it the "peaceful settlements of disputes, solidarity against aggression, reduced and controlled arsenals, and just treatment of all peoples."[5] Americans would finally get the "peace dividend," cuts in military spending to allow for increased government spending on domestic problems, that they had demanded since the fall of the Berlin Wall.[6]

From 1991 to 2001, a decade that I think is best described as a "decade of delusion,"[7] the United States would allow its irregular warfare capabilities to atrophy.[8] In the wake of our successes in Just Cause and Desert Storm, the United States deluded itself into believing that its conventional strength would be a sufficient deterrent in the post–Cold War era. The perception within the Department of Defense was that conflicts, both small and large, could be easily won by the newly formed U.S. joint force. Although this peerless high-end conventional capability proved a powerful deterrent to conventional adversaries, it would be of limited value and provided a false sense of security going

---

[3] George H. W. Bush, "Address to the United Nations General Assembly in New York City," September 21, 1992, transcript from Gerhard Peters and John T. Woolley, American Presidency Project.

[4] The end of the Cold War began in 1989 with the fall of the Berlin Wall and culminated with the collapse of the Soviet Union in 1991.

[5] George H. W. Bush, "Remarks at Maxwell Air Force Base War College in Montgomery, Alabama," April 13, 1991, transcript from Gerhard Peters and John T. Woolley, American Presidency Project.

[6] For example, in January 1990, three-fourths of Americans supported the idea of cutting military spending to allow for increased spending on domestic programs (Michael Oreskes, "Poll Finds U.S. Expects Peace Dividend," *New York Times*, January 25, 1990).

[7] John Gray, "Our Newest Protectorate," *The Guardian*, April 26, 1999.

[8] Steven Metz, *Rethinking Insurgency*, Carlisle, Pa.: Strategic Studies Institute, U.S. Army War College, 2007, p. 2.

into the population-centric conflicts that would dominate the years following the end of the Cold War.[9]

After a year at the Command and General Staff College, I spent the balance of the first half of this decade in Washington, D.C., first at the Army's Personnel Command and then at the Pentagon on the Joint Staff. During my first two years, as the Army's Special Forces field grade assignment officer, I learned firsthand that sustaining America's irregular warfare capability would not be a post–Cold War priority. In 1991, when I arrived, our Army special operations units were significantly understrength and struggling to man both the newly formed headquarters at Fort Bragg (USASOC) and Tampa (SOCOM). Yet Special Forces would be required to shed as much as the Army's overstrength branches, with total reductions in our ranks projected to be just under 30 percent.[10] Perhaps this was necessary for political reasons, but the outcome was a weakened American irregular warfare capability and reduced strength in the officer ranks among several year-groups, which would leave shortages in senior ranks that would last for the next 20 years.

However, during my next two years, now at the Joint Staff's Special Operations Directorate, I learned the continued value of irregular warfare in creating options for policymakers at the highest levels of government. For two national-level missions, I was asked to provide options that leveraged America's irregular warfare capabilities: (1) the attempted recovery of Lieutenant Commander Michael Speicher, the only U.S. servicemember missing in action from Operation Desert Storm and (2) possible U.S. support to Croatian forces fighting the Bosnian Serbs and their Serbian allies. In both cases, the irregular option was quickly rejected because the Department of Defense was skeptical of the risk. However, the United States would eventually employ such an irregular warfare approach in the Balkans by 1996.

---

[9] See, e.g., Cordesman, 2014.

[10] The plan, as of 1992, was for the Army active duty end strength to fall by 26 percent (Dwight D. Oland and David W. Hogan Jr., *Department of the Army Historical Summary: Fiscal Year 1992*, Washington, D.C.: U.S. Army Center of Military History, 2001, Chapter 7).

Operation Desert Storm: Validation of America's New Conventional and Raiding Capabilities

During the summer and fall of 1990, a small group of us at the Command and General Staff College gathered every few weeks for a notable war movie and beers. Together, we represented most of the combat arms, and the ostensible purpose of our "study group" was to discuss the issues that each movie highlighted. Perhaps indicative of the times, features included such classics as *Full Metal Jacket*, *Platoon*, and *Bridge on the River Kwai*. My selection was *Breaker Morant*, which for me is perhaps the best depiction of the ethics and dilemma of unconventional warfare, from both the tactical and national strategic levels.

Our group, which had honed its analytical skills discussing the profession of arms during these movie sessions (ably assisted by many beers),[11] turned the collective energy to analyzing options for invading Kuwait and Iraq in late 1990. In the end, the plan that we put together was close to what General Norman Schwarzkopf, who commanded coalition forces during Operation Desert Storm, would execute. Our armor and mechanized fellows were pretty clear on what the operational maneuver would require. As news reports came in about who was joining the coalition, I knew that our Special Forces teams would be assigned to the all Arab units, even the Syrian combat units, to help tie them into the theater-level plans by helping control their maneuver and fires. The only question we were left with was timing, but that too became predictable, as the logistics officer reminded us of the challenge of supporting this large a force in a "cocked" position. But it played out even faster than we had anticipated.

As good as we were, we had little sense of the long-term consequences that this short war would have on either America's post–Cold War view of how it should defend itself or (perhaps more importantly) the strategies of our adversaries. For the United States, the war clearly validated the potency of our conventional warfighting capabilities. But it also demonstrated the value of America's new raiding capability, with one observer describing the U.S. hunt for Scud missiles in the western

---

[11] We thought of it as our profession's version of *Mystery Science Theater 3000*.

deserts of Iraq as "conceivably . . . the most significant single operation of the war."[12] In contrast, America's irregular warfare capabilities provided limited results during the short war, as neither efforts to capitalize on Shia discontent nor the liaison role with non-NATO coalition partners played more than a supporting role in the invasion.[13]

In the following decade, the decade of delusion, the United States would prioritize resources for this new raiding capability and allow America's irregular warfare capability to atrophy. This prioritization reflected a general belief that the end of the Cold War meant the end of U.S. involvement in population-centric conflicts, a belief that was reinforced first by Operation Just Cause in Panama and now by Operation Desert Storm. As a result, irregular warfare "faded from the curricula of professional military education," and there was "little interest in developing new doctrine, operational concepts, or organizations" for this form of warfare.[14]

Operation Desert Storm also had a profound effect on U.S. adversaries across the globe. In addition to providing an impetus to modernize and develop indigenous precision-guided munitions capabilities,[15] the operation demonstrated the difficulty that these adversaries would face in competing directly against U.S. conventional military capabilities. This effect has been most clearly demonstrated for China, for whom Desert Storm was part of the impetus for both modernization and the development of "unrestricted warfare."[16] Unrestricted warfare called for the development and deployment of asymmetric capabilities in competition against the United States, using "whatever means are

---

[12] Adams, 1998, p. 234.

[13] Both of these efforts were implemented by U.S. Special Forces. The Special Forces Coalition Support Teams, who created partnerships with Arab militaries, were seen as somewhat more successful but still only a supporting effort for the priority mission (William M. Johnson, *U.S. Army Special Forces in Desert Shield/Desert Storm: How Significant an Impact*, Fort Leavenworth, Kan.: U.S. Army Command and General Staff College, 1996).

[14] Metz, 2007, p. 2.

[15] For a Russia example, see Defense Intelligence Agency, *Russia Military Power*, Washington, D.C., 2017.

[16] Andrew Scobell, David Lai, and Roy Kamphausen, *Chinese Lessons from Other Peoples' Wars*, Carlisle, Pa.: Strategic Studies Institute, U.S. Army War College, 2011.

at [China's] disposal, refusing to be fettered by rules and codes devised without its participation and which would work against it."[17] I am certain that it had a similar effect on the strategies of Iran and Russia, who would increasingly turn to asymmetric approaches in the years to come.

We might have been unable to predict the coming sea changes in either the U.S. national security establishment or the strategies of our adversaries, but the rich debate that we had during those beer and movie sessions proved a powerful analytical tool for unraveling some of the world's toughest problems. I would seek out or set up such a forum in subsequent commands, whenever feasible.

My Pentagon Wars: Fighting for America's Irregular Warfare Capability

In the summer of 1991, as the Soviet Union neared its collapse and my year at the Command and General Staff College came to a close, I headed to Washington, D.C., for my first post–Cold War assignment.[18] I was assigned as the Special Forces field grade assignment officer at then–U.S. Army Personnel Command, and it would be my job to collect the "peace dividend" for the taxpayer from the ranks of the U.S. Special Forces.[19]

My job, more unpleasant than I could have anticipated, was to hand pink slips to a full one-third of the U.S. Special Forces soldiers of my year group.[20] The United States was in a post–Cold War,

---

[17] June Teufel Dreyer, "People's Liberation Army Lessons from Foreign Conflicts: The Air War in Kosovo," in Andrew Scobell, David Lai, and Roy Kamphausen, eds., *Chinese Lessons from Other Peoples' Wars*, Carlisle, Pa.: Strategic Studies Institute, U.S. Army War College, 2011, p. 45.

[18] I had until then avoided being a staff officer at all costs, but staff work is the necessary evil of an American officer's career. The prevailing wisdom was to avoid it, stay in command as long as you can, and, whatever else you might do, stay away from Washington, D.C.

[19] I was the assignment officer responsible for promotable captains, majors, and lieutenant colonels for the Special Forces branch that was then two years old.

[20] Specifically, that was the 1978 year group.

post–Desert Storm era, and the Army would have to shrink substantially. However, while most branches found themselves with an excess inventory of officers, as the number of officer billets across the Army was slated to shrink by 25 percent,[21] the newly formed Special Forces branch (established in 1987) was still substantially understrength.

This was to be to the second major cut to the Special Forces ranks in under 20 years. In early 1979, when I entered the Special Forces Officer's Course, the Army had just finished a major reduction in the size of Special Forces, cutting total funding by nearly 90 percent and the force strength by 75 percent.[22] The resurgence in demand for Special Forces unique capabilities in the 1980s, particularly given President Reagan's proxy wars in Latin America, led to some modest increases in resourcing. But the consequence of this first round of cuts was that the Special Forces branch was significantly understrength when it was formed in 1987.

Although there were initially rumors that this post–Cold War reduction in force would be a "shaping exercise" and cuts to the Special Forces branch would be more limited, that was not to be. In the end, every branch would shed the same share of personnel, the logic being that we would not keep a bottom-third Special Forces officer in exchange for a bottom-half infantry soldier.

At the time, I was stunned by what I saw as an extraordinarily myopic decision by the Army, as it was culling its only irregular warfare capability during a time when our contests with these adversaries were on the rise. It was true that the conventional forces were taking equal cuts, and the pain of these cuts reverberated across the joint force.[23] But many of these branches were significantly overstrength (i.e., there were more officers assigned to the branch than dictated by the Army's

---

[21] The 1991 National Defense Authorization Act would require a 25 percent reduction in officers by 1995 (Pub. L. 101-510, National Defense Authorization Act for Fiscal Year 1991, November 5, 1990; Congressional Budget Office, *Reduction in the Army Officer Corps*, Washington, D.C., April 1992).

[22] Field Manuel 3-18, 2014, p. 1-9.

[23] Garry L. Thompson, *Army Downsizing Following World War I, World War II, Vietnam, and a Comparison to Recent Army Downsizing*, Fort Leavenworth, Kan.: U.S. Army Command and General Staff College, 2002, p. 59.

targeted end strength), while Special Forces was already understrength and would now be thinly staffed for a burgeoning set of requirements. This might have been acceptable if what was to come was a replay of the conventionally dominated Cold War era. Unfortunately, our enemies had other plans.

The most disappointing aspect of this assignment was the lack of support that the Special Forces branch received from the newly established SOCOM. It was rumored that SOCOM had been asked by the Army whether it wanted to protect our branch from the post–Desert Storm drawdown, and SOCOM had reportedly declined.[24] Whether true or not, SOCOM remained silent as the number of field grade officers in our ranks, as well as those of civil affairs and psychological operations, the other two key components of SOCOM's irregular warfare capability, was cut by a third.[25]

The reality was that current requirements rather than long-term capability development guided resource allocation in this post–Cold War period, and irregular warfare was no longer a national-level requirement for the Department of Defense. SOCOM had its hands full meeting the special reconnaissance and counterterrorism requirements being handed down to it from Washington, D.C., and many senior civilian and military leaders believed that the CIA should be America's irregular warfare force. In my mind, these requirements reflected outdated thinking and a failure to anticipate the national security challenges that would come. And I was not alone, as senior congressional leaders were similarly concerned that the Department of Defense was failing to adequately sustain its irregular warfare capabilities.[26]

---

[24] The story throughout the Special Forces ranks was that General Wayne Downing, the SOCOM commander, did not want to keep Special Forces officers who could not make the same standard in the regular Army. This was seemingly sound logic, but the sad truth is that many Special Forces officers were rated lower than their conventional counterparts because they were unorthodox, unconventional thinkers who did not do well in conventional units.

[25] The rumor mill was not kind to the senior Special Forces generals in either the Army or SOCOM at the time, who were blamed by the rank and file for not advocating on our behalf. I suspect—looking back—that they too were not consulted, as the senior Special Forces officer was only a two star and well below where corporate decisionmaking is done.

[26] Adams, 1998, pp. 205–206.

Without a higher headquarters willing to advocate for us, our office at Army Personnel Command, the proponent for the new Special Forces branch, led its own bureaucratic "unconventional warfare" campaign to build support for irregular warfare.[27] Our goal was very modest—specifically, to get support from SOCOM in forcing the Army to improve the command opportunity rates for our Special Forces branch, as we had much fewer battalion- and brigade-level command opportunities than other combat arms (and most other branches). Although seemingly a bit self-interested, the reality was that there were few colonels and fewer general officers in the Army who were expert in irregular warfare, and this was a strategic vulnerability that we sought to rectify.

The Special Forces branch was no stranger to advocating for itself when perceived as necessary for protecting the nation, and indeed the creation of the Special Forces branch was itself the result of its own years-long bureaucratic unconventional warfare campaign.[28] This previous campaign had included the establishment of the Special Operations Policy Advisory Group, reportedly built to overcome service resistance to institutionalizing special operations,[29] and a variety of staff officers who openly promoted "special-operations revitalization initiatives," often at the risk of their own careers.[30] In the end, even this change would likely have been impossible if not for the intervention of Army Chief of Staff General Edward (Shy) Meyer,[31] who understood that the Army must retain forces "capable of sustained operations

---

[27] Adams (1998, pp. 172–209) serendipitously uses the phrase "[s]pecial ops in the bureaucratic jungle" to describe the broader set of events during this time frame that led to both the revitalization of Special Forces (our focus) and the broader political struggle to establish SOCOM.

[28] The Secretary of the Army approved the establishment of the Special Forces branch on April 9, two weeks before SOCOM's establishment, with the branch established in mid-June 1987, some 35 years after the first soldiers were designated as Special Forces–qualified (Corson L. Hilton, *United States Army Special Forces: From a Decade of Development to a Sustained Future*, Carlisle, Pa.: U.S. Army War College, 1991, p. 11).

[29] Adams, 1998, p. 193.

[30] Adams, 1998, p. 188.

[31] Hilton, 1991, p. 5.

under the most severe conditions of the integrated battlefield . . . [and] equally comfortable with all the lesser shades of conflict,"[32] and he wanted to preserve the Special Forces for this unique role.[33]

During my tenure, from 1991 to 1993, our approach was largely unchanged from that used by our Special Forces brethren during the previous decade: We would influence the senior leaders in the chain of command by educating them ("teaching up") about the unique contributions that we provided the Army. We were supported by the two branch chiefs for whom I served, who understood that we were on an unconventional warfare mission within the Army personnel bureaucracy to influence and educate our peers and leaders.[34] And a friendship with a senior civilian leader at the Office of the Assistant Secretary of Defense for Special Operations/Low Intensity Conflict allowed us to insert language highlighting the problem and recommending solutions into one of its annual congressionally mandated reports.[35] Neither approach would bring any meaningful change, although command rates began to improve a decade later because of the excellent work of our young Special Forces officers on the battlefields after September 11.

Two years in that position were more than enough, as they were tough years of drawdown, uncertainty of mission, and growing resentment within the Army. Now, more than 30 years later, I still remember making those dreaded calls to my friends and recommending that they take the generous voluntary separation package.[36] But I got to

---

[32] General Edward C. Meyer, "The Challenge of Change," *Army 1981–82 Green Book*, October 1981, p. 14 (referenced in Hilton, 1991, pp. 5–6).

[33] Adams, 1998, p. 183.

[34] Our branch chiefs, only lieutenant colonels, had limited influence over the two-star commander of Army Personnel Command. I would come to understand only later that these decisions were made way above the level of even the two-star commander. He might have been in the room, but he did not make the decision.

[35] This was the office's annual personnel monitorship report.

[36] Banking on a shaping exercise, I had dialed up only half of what would be our quota in the salami slicing that eventually came. One officer, a West Point and Command and General Staff College classmate, asked whether he should be worried with his last report being in the lower half on a dual center of mass in a staff job in the special operations staff in Army

know every major and lieutenant colonel and half the colonels in Special Forces and came away proud of my new Special Forces regiment and the selfless patriots who took on the work of going into hostile and denied areas beside foreign partners.

This experience initiated what became a career-long tendency to reflexively defend my Special Forces branch. Only later would I realize that the resentment toward my branch that I saw then was a consequence of the widespread misunderstanding of the importance of irregular warfare. As events would soon prove, America's threats would transition to those for which special operations, particularly irregular warfare capacity, were best suited.

Developing Unconventional Warfare Campaigns: Creating Special Operations Policy Options

The second half of my four-year post–First Gulf War tour in Washington, D.C., was with the Joint Staff's Special Operations Directorate. Here I would learn that policymakers want options and that one of special operations' greatest contributions is in designing unique and innovative approaches that are molded to the requirements of the mission.

My primary focus during these two years was on solving the mystery of Lieutenant Commander Michael Speicher. Speicher became the first American combat casualty of Operation Desert Storm when his

---

Pacific headquarters. I said no, given he had been selected to resident command and staff, then a top 50 percent cut. He was selected to be separated, with a much smaller compensation package. I wrote letters of support to his congressman, confessing my screw-up to no avail. No luck with the inspector general either. Reversing an Army board decision was not going to happen—bad precedent. My guilt was compounded when I learned that he had a special needs child and his expenses were far from normal. But he tenaciously fought the decision and received a stay while his case was being reviewed by the Secretary of the Army. I was to learn an important lesson—the Army does not owe its soldiers a thing; it is those who make up the Army who owe each other. I received a call from the secretary's office: "Tell my friend to stop talking to the press and calling Congress—an early retirement program was going to be announced in a week and if he submits for early retirement, it will be approved." My friend retired four months later with a good Army pension and benefits.

F/A-18 Hornet was shot down on the first night of the invasion,[37] and he was the only American missing in action during the war.[38] We had never recovered Speicher's body, and it was a question as to whether he was actually captured by Saddam Hussein's forces and was being held in a prison someplace or whether he died on impact. His whereabouts were unknown until early 1993, when a Qatari hunting party smuggled a numbered part of his downed U.S. jet out of Iraq.[39] When we linked the serial number on the smuggled part to Speicher's aircraft, the search for Michael Speicher began.

This mission, which we called Operation Promise Kept, taught me the value of irregular warfare in creating options for policymakers. Ultimately, the chairman selected the plan that we developed at the Joint Staff, in which U.S. officers would overtly (e.g., in coordination with Hussein's government) accompany International Committee of the Red Cross representatives to the suspected crash site.[40] However, there was also a CIA plan, a SOCOM plan, and a 5th Special Forces plan, which provided, respectively, a civilian paramilitary, a unilateral, and a partnered military approach to policymakers. Each approach was viable, but each carried unique risks.

The mission was delayed six months because of "bureaucratic problems" within the Iraqi government, and members of the team would find the crash site already excavated when they arrived.[41] Advocates of the other three approaches, which were all covert and would have accessed the site without the Iraqi government even knowing the United States was interested in it, complained that the failure of the mission "reflected the agonies of an army trying to do its duty with-

---

[37] Tim Weiner, "With Iraq's O.K., a U.S. Team Seeks War Pilot's Body," *New York Times*, December 14, 1995.

[38] Thom Shanker, "U.S. Pilot's Remains Found in Iraq After 18 Years," *New York Times*, August 2, 2009.

[39] Qatar had become a U.S. ally in the wake of Operation Desert Storm, signing a defense cooperation agreement in 1992, and understood the importance of this information to us.

[40] Weiner, 1995.

[41] Tim Weiner, "Gulf War's First U.S. Casualty Leaves Lasting Trail of Mystery," *New York Times*, December 7, 1997.

out risking its soldiers."[42] However, the senior leaders in the Pentagon did not think that the covert option was worth the risk, as the site was already three years old at that point.[43] The mystery of Michael Speicher would remain unsolved until 2009, some 14 years after I left the Joint Staff, when U.S. marines in Anbar Province were directed to the site where Bedouins had buried his body 18 years before.[44]

Years later, General Wayne Downing (the SOCOM commander at the time of the planning) would comment that he wished he had asked the Jordanians to conduct the operation, in which Jordanian special operations forces would have executed the mission with covert U.S. support. This comment was particularly striking because SOCOM had resisted the development of this irregular warfare option at the time. My boss at the time, then-Colonel Jerry Boykin, who was a legendary figure in the national mission force community, understood the value of the irregular option. His support and the signal that its inclusion in the suite of choices would send was critical in forcing SOCOM to provide this option. I would not be surprised if the resistance to the irregular warfare option came from within SOCOM's staff, and not the senior leadership, as I would often find myself at odds with this staff over the next 20 years of my career.

My second major project while at the Special Operations Directorate was to develop unconventional warfare options for supporting Croatian forces fighting the Bosnian Serbs and their Serbian allies. Two of us, both Cold War 10th Special Forces veterans, worked together to develop and then brief a few options to the three-star director for strategy, plans, and policy at the Joint Staff. After the briefing, he tasked us to come back and give him what was essentially a tutorial on unconventional warfare. It was the first glimmer of hope that the field of work I had chosen indeed had relevance in the era of the new world order.

---

[42] Weiner, 1997.

[43] Weiner, 1997.

[44] Shanker, 2009. These marines were acting on intelligence provided by the Special Forces–led Combined Joint Special Operations Task Force–Arabian Peninsula (personal communication with Lieutenant General [retired] Ken Tovo).

In the end, none of the options that we proposed was chosen. Although what we were proposing was a popular notion with some in Congress, the memory of Vietnam was still strong, and senior Pentagon leaders strongly desired to avoid military involvement. But the fact that an unconventional warfare option was requested was evidence, at least to me, that senior leaders were beginning to recognize the value of irregular warfare in creating options for the United States. It meant that they would have to invest in the new capability, so that it would be available when needed.

CHAPTER SEVEN

Peacekeeping in Bosnia and the Reemergence of Irregular Warfare

On December 7, 1995, as luck and weather would have it, I led the first elements of NATO's newly established peacekeeping force into Bosnia. We were the front end of the 10th Special Forces contingent that would help enforce the peace agreement,[1] arriving in Bosnia the week before the Dayton Accords were officially signed.[2] The destruction that we saw when we arrived was breathtaking. There was something incon-

---

[1]   Colonel Geoffrey Lambert—the 10th Special Forces Group commander—had intended to lead the first elements in, but he had been in Vicenza, Italy, coordinating with the Army units that were coming into Multi-National Division (North) and got weathered out for almost a week. So, on December 7—a week before the peace agreement was formally signed by the two parties to the conflict—my small team of 14 special operators, who would become the special operations liaison officers to the three military divisions that would compose the NATO Implementation Force, was the first to deploy into Bosnia (from Brindisi, Italy). At the time, the United Nations was still in charge, and we were met by a U.S. Marine Corps officer who had been seconded to the United Nations. He wanted our names and social security numbers, and I remember telling him that we were the front end of a whole lot more coming and I doubted they'd be stopping to check in.

[2]   After landing at Sarajevo Airport, we were escorted by British special operations elements to Kiseljak, a resort town in Bosnia Croat territory on the outskirts of Sarajevo, where we would establish our special operations task force with the UK's Directorate for Special Forces. As I recall, we had to transit armed checkpoints run by all three ethnic groups to get there. Our compound was on the grounds of a resort hotel that was the assembly area for the retrograde of UN forces, and became a national-level Star Wars cantina of blue-helmeted military units. And it became utter pandemonium as some of our support forces began to arrive in Sarajevo—arriving U.S. units had to guard their equipment to make sure it was not stolen, as would happen to a Psychological Operations team that showed up to the hotel a couple of days later.

gruous about landmines in vineyards and children playing next to the burned remains of their former neighbor's house, a house quite possibly torched by their parents. War in developed countries stands out more, sectarian conflict especially—its ugliness more evident.

The Dayton Accords, officially signed by the parties to the conflict on December 14, 1995, ended a brutal and bloody sectarian war in Bosnia that had been ongoing since April 1992.[3] Within a week, a newly established NATO headquarters took over the Bosnia mission from a United Nations force that had proved unable to prevent escalating sectarian violence during its three-year mandate.[4] This NATO force would be responsible for ensuring the terms of the Dayton Accords (e.g., withdrawal of forces beyond agreed-to cease-fire lines) and was "authorized to operate under 'robust' rules of engagement" to do so.[5]

This NATO peacekeeping mission, anticipated at the outset to last no more than a single year,[6] would last until 2004. Initially, the mission was led by a British three-star general, and Bosnia was split into three geographic sectors, with an American, British, and French division responsible for each.[7] Although the geographic divisions would remain, the command transitioned to U.S. leadership after the first year, and the total size of the force fell by half, from nearly 60,000

---

[3]   The Dayton Accords, also known as the General Framework Agreement for Peace in Bosnia and Herzegovina, was a peace agreement achieved at Wright-Patterson Air Force Base in November 1995 and formally signed on December 14, 1995 (Organization for Security and Co-operation in Europe, "Dayton Peace Agreement," December 14, 1995).

[4]   The NATO peacekeeping force was known as NATO Implementation Force (IFOR), which took over the Bosnia mission from the United Nations Protection Force.

[5]   U.S. Army, Europe, "Military Operations: The U.S. Army in Bosnia and Herzegovina," Army in Europe Pamphlet 525-100, October 7, 2003, p. 15.

[6]   U.S. Army, Europe, 2003, p. 15.

[7]   The British-led three-star element was the Allied Command Europe Rapid Reaction Corps (ARRC), which led the ground component for the overall operation (Operation Joint Endeavor). The UK had responsibility for Multi-National Division (South-West), which was originally located at Gornji Vakuf-Uskoplje and later at Banja Luka—there was a great little bar at Banja Luka when I was there.

during 1995 to around 30,000 in 1996.[8] Further reductions in the size of the peacekeeping force would come in 1999, after the security situation had improved,[9] and an enduring European Union force would take over the mission in 2004.[10]

Special operations teams would play a critical role in maintaining the peace. During 1995–1996, national mission teams tracked down war criminals, 10th Special Forces facilitated the participation of non-NATO units in this peacekeeping mission, and British special operators were building networks of influence throughout Bosnian society as part of the unconventional warfare Joint Commission Observer mission. In December 1996, as the overall command of the Bosnia operation transitioned to the United States, 10th Special Forces would take over this unconventional warfare mission from the Brits. We would quickly learn that there was no term in the U.S. military vernacular or in U.S. military doctrine that adequately captured the activities involved in this mission.

The Joint Commission Observer mission would in many ways herald a rebirth of unconventional warfare in U.S. Special Forces, and lessons learned from Bosnia during 1995–1999 would remind us of the central role the indigenous population can play in conflict. This would underscore our approach in Afghanistan less than five years later. During the Bosnia operation, we demonstrated clearly how an irregular warfare capability could achieve strategic effects relevant to the main effort, in one case providing a mechanism that prevented rioting after a controversial political decision. Yet the U.S. Special Forces executing this nonstandard mission would have to justify themselves anew to each arriving U.S. conventional commander. Most of these senior leaders had spent their careers preparing to keep back the Soviet

---

[8]   The NATO Implementation Force would transition into the U.S.-led Stabilization Force (SFOR) in December 1996.

[9]   NATO Stabilisation Force, "History of the NATO-Led Stabilisation Force (SFOR) in Bosnia and Herzegovina," undated.

[10]   As of 2020, this enduring mission, Operation Althea, was alive and well, with some 600 forward-deployed troops provided by 20 troop-contributing countries (European Union, "EUFOR Operation ALTHEA European Union Military Operation Bosnia and Herzegovina," fact sheet, February 2020).

horde at the Fulda Gap and were simply not prepared to accept the value of our unconventional warfare mission in this population-centric conflict.

This Bosnian mission also demonstrated that the United States was not well organized to support an irregular warfare mission, such as the Joint Commission Observer. The difficulty that U.S. Special Forces faced in communicating our mission to U.S. conventional leadership was a striking example of this, particularly in contrast to the British special operators, who had led the mission from 1994 to 1996 and had excellent communication with their own senior leadership. But our deployment cycle (U.S. Special Forces battalions would deploy for only four-month rotations) created significant turbulence. It did not help that each new Special Forces unit's view of the mission varied significantly from that of the unit replacing it. This was again in sharp contrast to our British predecessors, who had a single commander responsible for the mission for the duration and managed their own rotations. Developing the institutional structures to support these types of missions would be a constant struggle throughout my career.

Rebirth of Unconventional Warfare: The Joint Commission Observer Mission

I deployed three times to Bosnia, first observing, later supporting, and finally leading the Joint Commission Observer mission. We provided the theater commander and the three division commanders ground truth about their respective areas and a mechanism for accessing and influencing key leaders throughout society. Although our nightly reports became valued sources of intelligence, our teams' ability to work established contacts at the local, regional, and national levels was most critical. This network, reinforced by the lethality that our Special Forces represented, acted as a means on the part of the coalition to continuously mitigate problems at a local or regional level that might otherwise evolve into ethnic cleansing or violence. These mitigating actions were often taken without explicit direction (or interference) from any higher headquarters.[11]

I would later refer to the Joint Commission Observer mission as a "contemporary" unconventional warfare campaign, as what we were doing did not fit nicely into our lexicon. Major Ken Tovo, then working for me as the company commander assigned to the U.S. division in northern Bosnia, described our mission to the U.S. division commander in his area as a type of *human intelligence* operation. It was a term the commander understood, and there was no question that we were generating tremendous volumes of intelligence, as our nightly operational report was more than 30 pages. But it was also inaccurate.[12] Our aim was to manipulate the local situation to the benefit of U.S. objectives, and good intelligence was simply a beneficial byproduct. In fact, we deliberately left out some of what we had learned, when it gave us leverage with local officials and when I feared it would prompt an unnecessary overreaction from the conventional and intelligence headquarters, claiming the turf where we operated.[13]

This Joint Commission Observer mission had originated in 1994, during the failed United Nations mission, when the British commander at the time had requested the support of "a few good Brits" to function as military emissaries. Their official duties would include escorting United Nations relief convoys, providing eyewitness accounts of key events, and, in at least one case, directing NATO air strikes. As

---

[11] It proved an excellent example of *mission command* well before the term became fashionable.

[12] Tovo would take away the same lesson that I did, and this experience was a precursor to the "inform and influence" strategic value proposition that he would later promulgate as the USASOC commander (personal communication with Lieutenant General [retired] Ken Tovo).

[13] If we put everything we knew about the criminal element in Brčko into our reporting, the conventional division's military police would have shown up and tried to disarm the guys that were guarding our contacts. And if we were doing intelligence operations, then there was a set of rules and requirements that would have changed the nature of our activity on the ground and greatly impeded our ability to mitigate the problems the teams were encountering on a regular basis. Frankly, those requirements would have been a huge impediment to what we were trying to do, as we knew things about our contacts that we could never put into our reporting.

I would learn, their mission was much broader than this, as they were building networks of influence throughout Bosnian society.

When 10th Special Forces first deployed into Bosnia in December 1995, this unconventional warfare mission was still British and would remain so until the end of 1996. The primary mission assigned to U.S. Special Forces was, instead, to facilitate the participation of non-NATO units in this peacekeeping mission.[14] Special Forces teams partnered with Czech, Egyptian, Hungarian, Malaysian, Polish, Romanian, Russian, and Turkish units distributed throughout the country, leveraging their languages, regional experience, and maturity to do so. The requirement for these liaison coordination elements, a concept that had been pioneered by Special Forces during the First Gulf War,[15] largely disappeared by the end of 1996.[16] Variants of this model would appear in a variety of subsequent conflicts.[17]

Although the focus of U.S. Special Forces during 1995–1996 was this liaison mission, it offered us the opportunity to watch (and learn) from what our British counterparts were up to. At the time, our 10th Special Forces contingent worked for a British-led special operations task force, which was commanded by a British general officer who was

---

[14] The mission was to "accompany a non-NATO unit in its duties, advise the non-NATO unit on NATO procedures, [and] provide NATO-compatible radios, English language ability, fire support, air support and medevac support as needed." D. Jonathan White, *Doctrine for Special Forces in Stability and Support Operations*, Fort Leavenworth, Kan.: School of Advanced Military Studies, Command and General Staff College, 2000, pp. 19–20.

[15] These were our Desert Storm coalition support teams.

[16] Although there were initially more than a dozen liaison coordination elements, all but three of them—those partnered with Hungarian, Romanian, and Russian units—had completed their mission by 1997. As of 2002, only a single liaison coordination element—that partnered with the Russians—remained (Bruce R. Swatek, *Role of Special Forces Liaison Elements in Future Multinational Operations*, Fort Leavenworth, Kan.: U.S. Army Command and General Staff College, 2002, p. 45).

[17] As an example, Special Forces teams in Afghanistan would partner and advise Jordanian and Emirati special operations teams that were themselves advising Afghan special operations units. The addition of a Muslim component to the coalition advisory work was an important advance in our approach to foreign internal defense or advisory work for obvious reasons. In Iraq, El Salvadoran conventional units would have Special Forces advisers alongside them, essentially fulfilling the same function they did in Desert Storm a dozen years or so prior.

also the commander of the newly formed Directorate of Special Forces (more commonly known as DSF). The 10th Special Forces commander played the role of deputy, and several 10th Special Forces staff officers (myself as the group executive officer included) were part of a kludge with Directorate of Special Forces staff officers and other augmentees that formed a combined special operations headquarters.[18]

We watched as our British counterparts leveraged their Joint Commission Observer mission to provide senior British commanders access to key leaders throughout society and an informal way of communicating with them. It required special operators out on the ground who could protect themselves and build an informal network of contacts to both gather intelligence and be in a position to mitigate problems locally. It was immediately clear that it was a true Special Forces mission.

Our initial exposure to the Joint Commission Observer would prove critical as, little did we know at that time, 10th Special Forces would take over the mission in December 1996. We found ourselves building out an intelligence and influence network and conducting, what the old-timers back in the 1950s had called, special forces operations. The Brits had been very secretive, but we knew that their special operators had national intelligence guys embedded with them and were running their own sources. They kept their sources, even after we took over their mission, and we found ourselves doing the same thing that they had been doing: recruiting, unofficially and officially, agents of influence to provide intelligence and be available to do things for us or to not do things, because we had our hooks in them.

From December 1997 to April 1998, my third deployment to Bosnia, I led this Joint Commission Observer mission as the commander of 3rd Battalion, 10th Special Forces. It was now my job to manage, with my battalion staff and my battalion's assets, units executing this very nonstandard mission in 14 locations spread across the country. It was a presence mission, which was sometimes difficult for our conventional division commanders to understand. Our mission

---

[18] This was the Combined Joint Special Operations Task Force (CJSOTF) that is discussed in more detail later in this chapter.

was not to raid or surveil the enemy; it was to create options using our indigenous networks for senior commanders and for the policymakers in Brussels and Washington.

I told my teams that their job was to have a network that allows them to know, influence, and act on problems that will inevitably occur in this very messy space. Each team was asked to build out a campaign for its four-month deployment based on a simple rhetorical question: Who are you going to have each of your 120 lunches with and why? We categorized individuals based on a construct that we called the *influence hierarchy*, which ranked individuals based on their affiliation (e.g., political, military, criminal) and level of influence (e.g., local, national).[19] The teams would project out whom they might have to influence in their assigned area of operation to affect an outcome or mitigate a problem and then would start identifying whom they were going to have lunch with. The company-level headquarters provided oversight, logistics, and communications support, while the battalion-level headquarters provided intelligence vetting, warehousing of information and intelligence, and policy overwatch.

As we got to work, we were surprised by the lack of what was (at least from our perspective) critical information about the key actors in the conflict. Our conventional headquarters and their intelligence agencies knew the serial number of every artillery piece in the cantonment areas but knew next to nothing about the local leaders, power brokers, and influencers. Basic information about mayors of prominent cities—e.g., names, political affiliations—was simply not available. Such information was apparently not a theater priority. This was remarkable, as these individuals were dealing with the day-to-day problems of keeping the peace among neighbors who had been killing each other just two years earlier. We would quickly remedy this problem in-house, because we would not be able to accomplish our mission without doing so.

---

[19] The categories were political, military, police, civic, media, nongovernmental organization, economy, crime, religion, and other (Charles T. Cleveland, *Command and Control of the Joint Commission Observer Program—U.S. Army Special Forces in Bosnia*, Carlisle, Pa.: U.S. Army War College, 2001, p. 12).

But we were also building an influence network, creating options for the senior commanders and for the policymakers. It was a long-term effort to be able to influence the people in this population-centric conflict, to build the capability for operations in what I would come to see as an emerging and underserved human domain of conflict. We co-opted the bus company presidents and got to know the criminal element, both of which had a hand in the violence and could be a powerful force in stopping future violence.

The potency of this influence network was demonstrated when the Brčko arbitration decision, which would determine the future of the city, was coming about.[20] The senior NATO leadership was concerned that this decision would again be accompanied by rioting and violence and needed tools to prevent this violence.[21] We were able to identify additional individuals, using our influence hierarchy, who likely played a role in the earlier riots. And we made a plan for our Special Forces teams to get to those individuals well in advance of the decision.[22] In short, we gave the senior commanders a mechanism to control the Brčko crime syndicate, black marketeers, and religious leaders to avert the potential violence.

Explaining just what we were doing to the U.S. division commander was the most difficult, because the term *unconventional warfare* had become heresy in the U.S. military, even in Special Forces. I recall being told to strike all references to *unconventional warfare* from briefing and other materials, as "this was something that we no longer did" after the end of the Cold War. But we were giving the division commanders the capability to apply influence and, if needed, coerce local

---

[20] Brčko was a disputed municipality and, as it was judged to be "too contentious to be resolved in the Dayton Peace Agreement," its final status was left to the Arbitral Tribunal to determine its fate—this decision was set for March 1998 when I arrived, although the decision would be postponed effectively indefinitely (International Crisis Group, *Brcko: What Bosnia Could Be*, Brussels, February 10, 1998).

[21] There were riots and violence just a few months before—in August 1997—related to the undetermined status of Brčko (see, e.g., United Nations, "Letter Dated 14 October 1997 from the Secretary-General Addressed to the President of the Security Council," S/1997/794, October 14, 1997).

[22] Cleveland, 2001, p. 14.

authorities to support our mission. What we were doing was a classic Special Forces operation. Interestingly, it was the French and UK commanders, who had either cut their teeth in the liberation wars in Africa (French) or had a background with special operations in Northern Ireland (UK), who best understood the value that we provided.

Communicating the Value of Irregular Warfare: Growing Pains in Achieving Unity of Effort

When 10th Special Forces arrived in Bosnia in December 1995, we fell under a UK special operations task force commanded by Cedric Delves,[23] a former commander of the UK's elite Special Air Service and a legend within the UK panoply of flags and leaders. Delves had a personal relationship with the three-star UK commander leading the overall mission, which meant that he was receiving guidance directly from the NATO commander. It helped that the three-star commander was himself a former commander of the Special Air Service and intimately understood its capabilities.

The relationship between special operations and the three-star NATO headquarters transformed radically when the theater command transitioned to the United States in November 1996. As the United States also took over the indigenous warfighting component of the special operations mission, command of the task force would fall to a Special Forces colonel from the reserves. He turned out to be a great American patriot and soldier, but the departure of the well-connected UK special operations leadership and national mission forces meant an immediate change in the access and credibility with the largely conventional U.S. higher headquarters.[24]

---

[23] The forces from the UK Directorate of Special Forces were deploying into country simultaneously, falling in on the UK special operations element that transitioned from the United Nations to the new NATO force.

[24] SOCOM's main effort was instead the capture of war criminals, which leveraged its commando wing. The decision to put a reservist in this role signaled the relatively low regard that SOCOM held for our indigenous warfighting effort.

I was the first deputy commander for this new U.S. Special Forces formation. My job, as the senior active duty guy, was to synchronize our efforts with the incoming American leadership. I had been instructed by the senior special operations commander in Europe (then–Brigadier General Geoff Lambert, the Special Operations Command Europe commander) to ensure that U.S. Special Forces would retain the freedom that we would need to execute the mission set that we were inheriting from our UK counterparts.

My moment of truth came just weeks after I arrived, when Major General Montgomery (Monty) Meigs, the new American division commander who had just arrived in theater, announced that he wanted unity of command among all U.S. forces in his division's pea patch in northern Bosnia. He wanted every U.S. citizen, and certainly everyone in a uniform, in the north to be working for him. For us, he wanted operational control of our Special Forces teams operating in the north, which meant that he and his staff could reorganize teams and assign missions as he saw fit. This was rumored to include dividing the Special Forces teams into any number of small elements that would then be assigned as liaison elements for his own forces.[25]

Meigs undoubtedly understood how our forces could be an asset for the strange environment that was now his battlefield. But the difficulty was that the operations that we were doing in his operational area directly affected our operations in the UK and French areas. Meigs had only one-third of the fight, and the bottom line was that any misuse of the force and its placement or unneeded risk assumed in his division's area in the north would have direct consequences for our ability to execute operations elsewhere—and vice versa. And if we lost our ability to conduct our Joint Commission Observer mission in the north, it would effectively destroy the entire program, as the ethnic boundaries in no way corresponded to division boundaries.

---

[25] The U.S. Army Europe commander had put an attached Special Forces battalion on Meigs's troop list. But this was only the newest "front" in what was an ongoing front between Lambert and conventional staff headquarters staffs over the command and control of Special Forces.

So, Brigadier General Lambert sent me to the new division head-quarters to talk to Major General Meigs. His guidance was simple: I was to tell this two-star commander, who was supported by the three-star U.S. Army Europe commander, that he was not going to get opera-tional control of our guys. The idea was simple enough, as it was clear that our battalion would be more effective in supporting his division if he left us alone and under the operational control of the three-star headquarters.[26] But it fell to me to make Meigs like the decision. At the time, Lambert told me that "sometimes you have to put a good man between you and the problem." As I had just come out on the battalion command list, I was to be that sacrificial "good man."

In preparation for what promised to be a contentious meeting, I convened a small planning group that included the current Special Forces company commander operating in Meigs's region in the north and Major Bennet Sacolick, a former subordinate who had been a Spe-cial Forces team leader for me in Panama, who just so happened to be in Bosnia with a U.S. special mission unit supporting war-criminal capture operations.[27] This talented and eclectic group developed a good plan for how our headquarters and our teams would support the American division commander. The briefing itself completely avoided a discussion of the sensitive command and control issue. But I put it in the read-ahead paper as frankly as possible. It said something along the lines of: "U.S. Special Forces will not pass operational control to the division because (1) there is nobody in the division staff who under-stands the Special Forces operations that we are undertaking, and (2) these operations have an effect on the adjoining division areas that are not under the control of the U.S. division commander."[28]

---

[26] The organization that we were proposing to Meigs was analogous to the relationship that the British special operators had had previously under the British-led three-star Allied Com-mand Europe Rapid Reaction Corps.

[27] Sacolick had joined the national mission force after his time in Panama.

[28] The first argument may have been a bit overstated, as Multi-National Division (North) did have a Special Forces major working in the G-5 (Michael L. Findlay, *Special Forces Inte-gration with Multinational Division–North in Bosnia-Herzegovina*, Fort Leavenworth, Kan.: School of Advanced Military Studies, U.S. Command and General Staff College, 1998, p. 26).

Meigs was a tremendous gentleman. He welcomed me and had dinner with me the night I arrived. That night, he told me that he needed intelligence but also a tactile sense of what the heck was going on. He needed *Fingerspitzengefühl* (literally, "fingertip feeling") of what was going on in the north.

The briefing went fine, and it was clear that he understood our mission and our approach. But then he pressed me on the command and control issue, indicating that my briefing had skirted the "key issue." I told him that it was in the paper, and so he read it while I waited. When he got to the discussion of the command and control, he concluded brusquely: "Well, then, you need a different division commander." He did not get emotional about it, as he knew well that this was the special operations community's position. My only response to him was, "Sir, this is something that you need to talk to General Schoomaker about."

Peter Schoomaker, then the three-star commander at USASOC, arrived a week later, and the problem disappeared. Schoomaker understood instinctually that what we were doing should not be put in a position to be undone by one-third of the tactical fight. And he defanged the whole argument with the first sentence out of his mouth: "Monty, your problem has nothing to do with the operational control of these forces." He explained Bosnia using the idea of a Rubik's Cube—how, if you turn one side, you turn a different face on the other side. Effectively, the problem disappeared because a senior special operations leader was able to take the air out of the argument.[29]

Schoomaker was not just the USASOC commander but a former Delta commander and assistant division commander at 1st Cavalry Division. He had a hugely outsized reputation within the Army lead-

---

[29] He was really masterful, in my view, although I must admit I wasn't concentrating as much on the argument because the USASOC commander had pocketed my last tin of stateside Copenhagen. When he first arrived, we met him up in the Multi-National Division (North) area and briefed him in a secure facility. The first thing that Schoomaker did was ask, "Hey, Charlie, you got any dip on you?" I did, and when I gave him my dip can, he took a dip and kept my can. As we were getting ready to leave, the briefing being over, my big issue was whether I should ask him for my tin back, which I did. I said, "Hey, sir, I want my tin back," and he had a smile on his face and gave me back my tin. Maybe it was a test.

ership constellation and would soon command SOCOM and later become the chief of staff of the Army.[30] But it required his intervention to counter the operational plans for Bosnia developed by U.S. Army Europe.[31]

The central challenge, which was not really the fault of the conventional planners at U.S. Army Europe, was an institutional failure to recognize this as a form of irregular warfare, as well as the absence of universally understood and accepted doctrine for irregular warfare campaigns. This was a challenge that would not be solved anytime soon, and although this was the first time that I was part of this debate, I would witness a similar debate over conventional ownership of Special Forces in every subsequent conflict in which I was involved.[32]

Organizing for Irregular Warfare: Growing Pains in Establishing the New Combined Joint Special Operations Task Force

In December 1995, I became the vice chief of staff for NATO's new special operations task force in Bosnia, which would be a mash-up of U.S. Special Forces and the then-new UK Directorate of Special Forces. This newly forming Combined Joint Special Operations Task Force (CJSOTF) was still a relatively new operational construct. Goldwater-Nichols was less than ten years old at that point, so operational constructs with a *J* (for *joint*) were just starting to pop up. But with a *C* (for *combined*) in front of it, we were in uncharted territory. This CJSOTF would be the first operational employment of a NATO coalition special operations headquarters. Add to that the fact that the

---

[30] General Schoomaker retired after commanding SOCOM but was recalled to active duty to become the chief of staff of the Army.

[31] The one-star commander of a region's theater special operations command might be able to help, but that influence would wax and wane with the personalities and backgrounds of the four-star leadership.

[32] An interesting historical footnote is that I never saw the same debate over the U.S. national mission force or even the Army Rangers.

United States was not the lead nation, and it meant that we had our work cut out for us.

The UK and French forces inside the new CJSOTF, already established because they had been part of the previous United Nations mission, were very welcoming to their American counterparts infiltrating from Italy.[33] For one, they proved critical in helping me accomplish my first task: They took one of my Special Forces liaison elements directly to the UK division headquarters and coordinated access for one of my other liaison elements to the French division headquarters in downtown Sarajevo.[34] The Brits on the CJSOTF staff also hosted a wonderful Christmas bash. We, not to be outdone by our UK counterparts, had to reinfiltrate into Italy on a UK special operations aircraft to buy wine, cheese, and meats for a New Year's event. Amid the dystopia that was Bosnia that winter, these diversions were important relationship-building exercises. There are few fireworks as impressive as the thousands of tracer rounds and flares fired during celebrations in war zones.

But the desired outcome of a merged staff proved elusive during the CJSOTF's first year.[35] One challenge was that we were under a major time constraint: We had arrived in Bosnia and were immediately expected to perform as a staff. The expectation was that, as professionals, we could build a functioning operational combined headquarters as soon as we hit the ground. With American teams deploying to link up with coalition units and the Brits already engaged in Special

---

[33] I deployed from southern Italy with 13 or 14 special operators of various stripes, communicators and young Special Forces officers. I had three liaison officer packages—I sent three guys to the U.S. division, three as liaisons to the French division, and three up north—and the U.S. contingent for the CJSOTF staff.

[34] The UK division headquarters was at Gornji Vakuf-Uskoplje. Our new UK special operations partners provided introductions for the CJSOTF liaison team. Another UK team took the U.S. division liaison team to its base on the mine-infested Tuzla airfield to await the U.S. division's arrival. The U.S. division experienced challenges in crossing the Sava River. Our liaison element designated for the French division hopped a ride on a United Nations unit's old M113 armored personnel carrier that happened to be heading from Kiseljak, where the CJSOTF was based at the time, to downtown Sarajevo.

[35] An additional challenge was that the UK special operations teams resented the establishment of an in-country higher headquarters, as they had operated independently for more than a year.

Forces–like operations, there was no time for a staff exercise or even for the American and UK components of the CJSOTF to meet before the deployment. Our boss met their boss and an agreement was hashed out, with the 10th Special Forces staff elements essentially serving as a shadow subordinate staff to the UK Directorate of Special Forces principal.

What started as an ideal of a merged staff quickly evolved into two national contingents that worked to support each other but remained under essentially national control. Looking back, my relationship with the UK chief of staff, notionally my immediate supervisor on this combined staff, was less than perfect, and truth be told I probably was as prickly as he was. Protective of my staff, I could not stand it when UK officers chewed out my troops, even if for apparently good reason. That and a diminishing workload after the CJSOTF's move to the Ilidža complex in Sarajevo explain why I stayed only a few months in theater.

Building coherence among the various national special operations efforts would remain a persistent difficulty throughout this Bosnia mission. When I returned to Bosnia in October 1996 as the deputy commander of the CJSOTF, there was still no premission exercise to work staff functions, no manual detailing what each staff officer did, and no staff process yet established.[36] We would form an international staff with an excellent French chief of staff,[37] but the quality of the NATO personnel staffing our formation would vary. It did not help that the NATO nations, particularly the United States, resisted transferring any directing authority over their special operations forces outside their national chains of command.

Added to this was the reality that U.S. Special Forces were wholly unprepared for this type of operational-level requirement. We criticized (at least privately) our NATO partners for providing what we

---

[36] This would be my first exposure to a joint manning document staff. Although the United States was the framework nation, and 10th Special Forces would provide the bulk of the staff, our staff would be filled out by augmentees from other NATO nations.

[37] By now, the CJSOTF included significant contributions from the Dutch and French, in addition to ourselves and the Brits.

called a "dog's breakfast of augmentees" to the U.S.-led CJSOTF, but we did not do much better. It was telling that our initial staffing included a bunch of unidentified lieutenant colonels and colonels with a reservist as the initial commander. We would correct this, in part, when Les Fuller, the sitting 10th Special Forces commander, deployed with a handful of our best staff officers as the CJSOTF commander in March 1997. But we would continue to cobble together the CJSOTFs piecemeal.[38]

We were also not structured to provide continuity across the very nonstandard, long-term irregular warfare mission that we were now responsible for. The Joint Commission Observer mission required that we build relationships with assets and foreign units, which was complicated by our four-month deployment cycle. The turbulence created by our relatively short rotations was exacerbated because each unit's view of the mission varied significantly from that of the unit replacing it. This turbulence occurred even when the mission was handed between two different 10th Special Forces battalions,[39] although it was worse when a battalion from 3rd Special Forces was added to the rotation.

Our approach was in sharp contrast to the UK's, as a single UK special operations unit (the Special Air Service) had established a presence, had managed its own rotations, and the commander of the unit never lost responsibility for the effort. We, in contrast, would hamper our indigenous warfare special operations efforts by rapidly rotating in new units and new commanders. This rotation policy is perhaps well suited for conventional forces, who are responsible for occupying territory. But it was crippling to our units, who must develop, maintain, and use a deep understanding of the conflict at the human level to

---

[38] The groups were everything. And the problem with deploying a Special Forces group to a mud spot such as Bosnia is that the rest of the theater gets neglected, unless 10th Special Forces was going to run all its operations for its entire theater from Bosnia. That might have been a better alternative, although it would have required augmentation that could not be sustained with the then-available special operations force structure.

[39] I saw this clearly at the end of my first tour, as Tom Rendall, who deployed as the commander of 2/10th Special Forces Group (Airborne) following the 3/10th Special Forces Group (Airborne) rotation, simply had a different approach for this new Joint Commission Observer mission.

provide nuanced, timely, and local solutions, whether kinetic, informational, or civic. We knew the problem at the time but did not have the institutional structures in place to consider other solutions, such as that used by our UK counterparts.

On top of all these challenges, we had to move the CJSOTF's headquarters. We moved from an old resort hotel in Kiseljak to an even older, mine-infested but beautiful resort complex on the Roman baths at Ilidža, a suburb of Sarajevo. Ilidža at the time was occupied by Bosnian Serb militia units, with whom we negotiated a relief in place. Strange times, but what a very interesting operating environment. Just dangerous enough that you needed to remain armed but permissive enough that you could go anywhere if you needed to.

Coalition warfare would prove to be a central component of our irregular warfare efforts, and broader U.S. national security strategy, in the 20 years following Bosnia. What we learned at the CJSOTF in Bosnia was that the aggregation of headquarters, forces, and staff requires flexibility, creativity, and practice. It requires a flexibility to break down standard structures, the creativity to align the right capability against a given requirement irrespective of nationality, and practice to ensure that the aggregation is as natural as possible.

More importantly, we learned on the ground that the unconventional warfare skill set of U.S. Special Forces had an important post–Cold War application but that the hierarchy of the Department of Defense was ill designed to effectively leverage this irregular warfare capability. Little did we know at the time that unconventional warfare would soon become the primary effort in Afghanistan.

CHAPTER EIGHT

Unconventional Warfare in the War on Terror

In the summer of 2001—now a colonel and having just spent a year at the Army War College thinking and writing about war and peace and musing about how irregular war and my own experiences were often between the two—I took command of 10th Special Forces. I began this new role with a determination to revitalize the group's capability to conduct unconventional warfare in Europe. So that summer we selected, assessed, and trained a small group of Special Forces operators and organized them into teams designed to deploy behind enemy lines and in the most hostile, denied environments. Little did I know that they would see duty not in Europe but in first Afghanistan and then Iraq.

The United States was attacked that fall, on September 11, 2001. At the time, I was en route to a Hungarian Air Force base for a combined exercise (involving U.S. Special Forces and a U.S. special mission unit), and we listened in via radio as the events unfolded in New York and Washington, D.C. By the time we arrived at the site, the special mission unit was packing up to return to the United States. Three days later, returning to home station to stand by for what we all knew were orders to come, we silently watched the smoke still rising from the World Trade Center from 30,000 feet above.

The United States' response to the September 11 attacks would rely on its irregular warfare capability, particularly its unconventional warfare capability, and it was all hands on deck within the civilian and military special operations communities. The main effort was 5th Special Forces, and its 55 soldiers who deployed into Afghanistan would eventually help the Northern Alliance overthrow the Taliban. But 10th Special Forces supported these efforts by giving up our deputy commander, Mark Rosengard, a charismatic and supremely competent officer who would play a critical role in the successful toppling of the Taliban,[1] and sending a handful of our operators (the military's best mountain climbers) to help prepare the 5th Special Forces teams for operations in the Hindu Kush of Afghanistan.[2] And the unconventional warfare teams I had established were sent to work with the interagency to support sensitive operations.

Meanwhile, the rest of 10th Special Forces remained focused on operations in Europe: winding down operations in Kosovo, initiating a train-and-equip program in Georgia,[3] and redesigning and implementing a Europe-based group unconventional warfare exercise. In Kosovo, what had previously been the main effort, we revalidated the importance of building indigenous networks of agents of influence, lessons relearned and refined in Bosnia in the mid-1990s. Our efforts in Georgia and our exercise would leave us well prepared for what was coming next, a historic assignment to lead the blended special operations and conventional fight alongside the Kurds in northern Iraq.

Although September 11 framed my first year in command of 10th Special Forces, the invasion in Iraq was to dominate the second. In the late spring of 2002, we got wind of the possibility that the group would be part of an operation in Iraq. The whole notion seemed a bit far-fetched to me, but 10th Special Forces had a good relationship with the Iraqi Kurds that dated to the early 1990s and would be the logical choice to mobilize Peshmerga fighters against Saddam Hussein.[4]

---

[1] Capitalizing on this important work with powerful warlords, the "Lion of Judah" (as Mark would be called by his 5th Special Forces compatriots) would be appointed the first U.S. security assistance officer in the days following the Taliban's downfall.

[2] Within 10th Special Forces are the Special Forces' high-altitude and cold-weather experts.

[3] This mission would be transitioned to the Marine Corps in 2003 (Hope Hodge Seck, "Marines' Alliance with Georgians Holds Clues to Future Missions," *Marine Corps Times*, December 11, 2014).

[4] At the close of the First Gulf War, 10th Special Forces played a prominent role in protecting and providing humanitarian assistance to the Kurds, as part of Operation Provide Comfort. We were therefore the logical choice to harness the 65,000 Peshmerga fighters arrayed

Less than one year later, I would find myself commanding Task Force Viking, the multinational task force responsible for the northern front at the very beginning of Operation Iraqi Freedom.

The success of first 5th Special Forces in Afghanistan (in 2001) and then 10th Special Forces in northern Iraq (in 2003) demonstrated the efficacy of the Special Forces groups as a tactical headquarters for irregular warfare. Both groups demonstrated prowess in supporting and directing indigenous mass, fires, maneuver, and intelligence and proved their exceptional ability to adapt to the challenges of these new wars.

Yet those tactical successes would be insufficient to achieve enduring success in either conflict, in large part because the United States had yet to recognize that these wars were sufficiently different (from the wars for which it had built its military) as to require new doctrine, leaders, and organizations. The most glaring gap, in my view, was the lack of a literate irregular warfare headquarters above the Special Forces group. The consequence, in both cases, was that the United States was not organized or prepared to transition our initial tactical successes into an enduring strategic success. This remains an enduring challenge, across the myriad other population-centric conflicts that fester throughout the globe.

Building a Tactical Unconventional Warfare Capability for the 21st Century: 10th Special Forces Jedburgh Teams

The call for Special Forces to support the interagency came soon after the September 11 attacks, and we responded by sending two six-man unconventional warfare teams that 10th Special Forces had developed just that summer. These teams, patterned after the World War II Jedburgh teams that had deployed behind enemy lines to support the Normandy landings,[5] provided exceptional service in Afghanistan under a

---

along the 360-kilometer Green Line that divided the Iraqi Kurds and Saddam's army since the end of the First Gulf War.

5   CIA, "Surprise, Kill, Vanish: The Legend of the Jedburghs," December 17, 2015.

different authority. We would link up with these men again in northern Iraq in March 2003, where they would be part of the interagency vanguard supporting 10th Special Forces' lead role in the northern offensive of Operation Iraqi Freedom.

I established these Jedburgh teams in the summer of 2001, shortly after taking command of 10th Special Forces. My intent was to cement the group's capability to conduct unconventional warfare in Europe, a core responsibility of 10th Special Forces from my perspective, by investing in a small number of operators to prepare them for what would be the most hostile denied environments. We selected, assessed, and trained Special Forces operators for two six-man teams, with one team drawn from each of 10th Special Forces' two U.S.-based battalions.[6] Selection was restricted to only those operators who had attended special schooling in intelligence tradecraft skills, had shown exceptional judgment and maturity, and had fluency in at least one European language.

There was considerable pushback from across 10th Special Forces about the idea of standing up special teams, particularly from some of the old hands, who thought that the same standard ought to be applied to all teams and that consolidating talent could hurt the 12-man teams from which these Jedburgh teams were selected. This consolidation of talent was potentially problematic because Special Forces was suffering from a recruiting crisis at the time, and we were already short personnel. That was true, but the reality was that there was no way that the budget could support the training required for everyone, nor could the calendar allow the time away from other commitments. And the group needed to be sure that it could credibly operate in the most-difficult internal security circumstances.

The value of the unconventional warfare capability embodied in these Jedburgh teams was validated, at least to me, when I debriefed members of both teams just returning from Afghanistan over beers in spring 2002. They were clearly loving life. Despite not being eager to

---

[6]   At the time, each Special Forces company had a ghost team because of the chronic personnel shortages in the Special Forces branch. We directed the U.S.-based 2nd and 3rd Battalions to reactivate half of one of their three empty teams for these new Jedburghs.

join "my" Jedburgh teams, which they had thought a pipedream of their new commander, they were happy to admit to me that they had been wrong. Surprised in how quickly they were deployed into combat, they found themselves doing exactly what they had signed up for Special Forces to do. It was an opportunity for these soldiers to deepen their tradecraft and advanced special operations skills and to cement a relationship between Special Forces and other government agencies that would pay dividends for the United States in years to come.

We deployed our Jedburghs again in October 2002, this time to northern Iraq in preparation for 10th Special Forces' role in the imminent invasion. They were assigned to the Northern Iraq Liaison Element, an element from another government agency that had a long working relationship with the Kurds, and the resources, willingness to help, and ingenuity of this hybrid element proved a critical resource in establishing rapport with our Kurdish partners well in advance of the invasion.

This hybrid element facilitated our infiltration of two tranches of 10th Special Forces operators into Iraq in advance of the invasion. The first was a group of 30 Special Forces soldiers, which we called the "dirty 30," who had been part of the escort for a U.S. diplomatic mission into the Kurdish Autonomous Zone in February 2003.[7] This advance element allowed 10th Special Forces to mobilize the indigenous partner force of some 65,000 Kurdish Peshmerga fighters weeks in advance of the invasion. A few days later, this interagency element infiltrated two of my battalion commanders along with their opera-

---

[7]   U.S. Ambassador Zalmay Khalilzad, President George W. Bush's envoy to the Iraqi opposition, required a personnel security detachment for a diplomatic mission into the Kurdish Autonomous Zone in northern Iraq. Seizing on the opportunity to get more Special Forces soldiers into the area, 10th Special Forces gained approval from U.S. European Command and U.S. Central Command for 30 soldiers in the initial escort to remain behind to augment the Jedburghs and interagency element already in place. The plan, which involved entering and leaving by ground through the Habur border gate, called for the 30 soldiers and their vehicles to stage in a concealed position near the border crossing when the ambassador returned from his meetings with the Kurdish leadership. If the Turks gave the ambassador any trouble for being light on his returning security detachment, the 30 soldiers would move to link up with him. For whatever reason, the Turks registered no complaints, and the ambassador passed through without issue, leaving the 30 behind.

tions officers.[8] These battalion commanders, each a veteran of 10th Special Forces' mission to protect the Kurds in northern Iraq after Operation Desert Storm, linked up with the two Kurdish factions to which they were assigned.[9] Initially dressed in civilian clothing, each presented himself to the leadership of his faction in uniform when the invasion kicked off, a powerful and overt sign of U.S. commitment.

Our Jedburgh experiment, at least from my perspective, paid off in both Afghanistan and Iraq. However, I faced resistance throughout my implementation of this initiative, even within the irregular warfare–centric mind-set of the 10th Special Forces Group. The experiment demonstrated to me the criticality of commander vision and perseverance—but also the importance of communicating the reason for the change. In a widely misunderstood field, such as irregular warfare, you can never assume shared goals, outlook, and understanding. This experience would prove valuable in the years to come, as I would face increasingly bureaucratic headwinds in efforts to improve my small portion of America's irregular warfare capability.

Task Force Viking and Unconventional Warfare in Northern Iraq

In March 2003, 10th Special Forces led the northern offensive of Operation Iraqi Freedom. We were to be an economy-of-force mission, with my brigade-sized element responsible for preventing 13 northern-based Iraqi divisions from interfering with the main U.S. assault against

---

[8]   In the wake of the "dirty 30" episode, an offer came down the chain that we could deploy four more soldiers into northern Iraq. One of my bright staff officers—then–Lieutenant Colonel Darsie Rogers—recommended sending in the two battalion commanders, and they then selected their battalion operations officers to deploy with them. It was one of the best decisions I made in command. Although it took them away from their commands, it put them in the fight and allowed them to refine the plan that they would eventually execute. The information coming back to me was exactly what I and their troops needed.

[9]   The two factions were the Patriotic Union of Kurdistan and the Kurdistan Democratic Party.

Baghdad.[10] This was to be a true unconventional warfare mission, as it would rely on a partnership with the 65,000-strong Kurdish Peshmerga. It was also a mission that we were uniquely qualified to do, as 10th Special Forces had "saved a half million Kurds from extinction" just a decade earlier, and the Peshmerga were eager to work with us again.[11]

We first learned of the invasion and the role that 10th Special Forces would play in spring 2002, while conducting (in Germany) the first group-level unconventional warfare exercise since the fall of the Berlin Wall.[12] It was during this exercise that the commander of the Air Force's 352nd Special Operations Group, our sister organization in the European theater, agreed to merge his staff with 10th Special Forces (as we had done during this exercise) if a mission in northern Iraq came to pass. Although initially skeptical that we would be tagged to serve outside Europe, we established Task Force Viking at 10th Special Forces' group headquarters in July and began preparing our teams for a possible invasion.[13]

---

[10] Isaac J. Peltier, *Surrogate Warfare: The Role of U.S. Army Special Forces*, Fort Leavenworth, Kan.: School of Advanced Military Studies, U.S. Army Command and General Staff College, 2005, p. 24.

[11] The quote is referencing Operation Provide Comfort and is attributed to General John Galvin (USASOC, "10th SFG(A) History," webpage, undated-a). Two of my battalion commanders—Bob Waltemeyer and Ken Tovo—had been part of this previous mission, as had many of our senior noncommissioned officers. These relationships would allow us to rapidly and effectively mobilize a 65,000-strong Kurdish Peshmerga force for unconventional warfare.

[12] This was a revived Cold War exercise, Flintlock, which has since become a Special Operations Command Africa multinational exercise.

[13] Teams were isolated for planning and limited rehearsals, with commanders taking "brief backs" to ensure that they were prepared. In one particularly memorable briefing, a senior Special Forces soldier proudly described how he could just manage to squeeze his team's equipment into three Land Rovers but still recommended buying trailers to augment their load capacity. As events played out, we entered Iraq pretty much walking out of the back of an MC-130 Combat Talon with what we could carry in our rucks. Our Land Rovers would eventually catch up with us—thanks to the brazen work of my executive officer—but our vehicles were, initially, what could be bought by our Kurdish partners from their own citizens. My command vehicle was a well-used blue Toyota Land Cruiser with crushed blue

Despite our skepticism, planning for the invasion continued throughout the fall, and 10th Special Forces remained included in planning conferences that were increasing in frequency and length.[14] The operational pace was quick, and the planning effort very decentralized, a pace of decisionmaking that reflected a theater at war and a confidence in subordinate leaders that were exhilarating.[15] Although I had seen it before in Panama and the Balkans, it was amazing to see again what the United States could bring to bear when it decided that war was the likely policy option.

In mid-January 2003, I deployed with Task Force Viking's leadership and planning teams for a ten-day planning conference in Stuttgart but would not return to Fort Carson until the end of May (and via Iraq). It was at this planning conference that we learned that Task Force Viking would be the primary effort in northern Iraq if the 4th Infantry Division would not be permitted to pass through Turkey. Negotiations with Turkey had been ongoing since early 2003, with both U.S. Central Command (CENTCOM) and U.S. European Command negotiating for base access for us and overland access for the 4th Infantry Division.[16] The chief of the Turkish General Staff eventually granted

---

velvet velour on the dash and a Turkish evil eye dangling from the rearview mirror. It ran well.

[14] These conferences were hosted by U.S. Central Command and Special Operations Command Central.

[15] This was necessary for the one-star Special Operations Command Central that was still supporting combat operations in an uneasy Afghanistan while ramping up combat and advisory operations in the troubled Middle East. The cultural differences between "our" theater special operations command—10th Special Forces was historically aligned with Europe—and SOCCENT was on display and stark. Special Operations Command Europe, the grand dame of theater special operations commands, retained the feel of the Cold War and whiffs of World War II and the Office of Strategic Services. The pace was deliberate, action bound by alliance ties, and planning was dominated by the conventional military. SOCCENT, in contrast, had the year before supported the Afghan Northern Alliance in its overthrow of the Taliban government through an artful combination of Special Forces, the CIA, and American airpower. The impression for us outsiders was that removing the tyrant Saddam Hussein was another of many tasks on SOCCENT's to do list.

[16] The Turks also had a military presence in northern Iraq, and it was important that we keep them on our side—but also that they not interfere, as their suspicion of the Kurds and overplayed patronage of the Turkomens made them a potential problem.

Task Force Viking permission to operate from Diyarbakir, which offered a military airport and robust road access to northern Iraq,[17] but the Turks continued to slow-roll access for the 4th Infantry Division.[18] Once the U.S. invasion started, they would do the same to us.

In the end, Task Force Viking would take lead for the northern component of the U.S. invasion, staging in Romania for the operation after it became clear that the Turkish offer of Diyarbakir was disingenuous.[19] The 173rd Airborne Infantry Brigade, the Air Force's 352nd Special Operations Group, the 26th Marine Expeditionary Unit, and several other elements would eventually be either assigned or attached to Task Force Viking for the mission.[20] Subordinating a conventional brigade to our brigade-sized Special Forces task force was

---

[17] I'll never know precisely why but, at the end of the meetings, the Chief of the Turkish General Staff sought out our Special Forces delegation to ask whether we were happy with his decision. The Turks, it must be said, were extremely resistant to Diyarbakir because it was the unacknowledged Kurdish capital and instead had offered us a military facility at Batman.

[18] The Turks might have already decided that no Americans would operate from Turkey, so that the final outcome of the debates did not matter.

[19] In February 2003, when it became clear to me that the Turks were stalling, Task Force Viking began to assemble at an old Warsaw Pact air base in Constanta, Romania, that had originally been surveyed by Special Operations Command Europe as a potential exercise location. The U.S. European Command staff and U.S. Embassy quickly gained the necessary approvals, and the force and its growing support structure began assembling on the bitter-cold shores of the Black Sea in early March. The speed with which supporting agencies and commands sent assets was remarkable, with exotic communications, intelligence, and other technological assets appearing at the airfield daily. It was a push system, and it became a challenge to organize the free help. This was a good problem to have, but it did cause us to create an ad hoc battalion to take charge of integration, care, and feeding.

[20] At the planning conference, the deputy commander of the invasion's Combined Joint Forces Land Component Command asked us how Task Force Viking might employ the 173rd Airborne Infantry Brigade. I outlined a concept for him in which the brigade would relieve one of our battalions, and their 6,000-strong brigade of Kurdish irregulars, in securing the oil-rich region surrounding the culturally diverse city of Kirkuk, and he quickly agreed. Ultimately, we would have a force of more than 70,000 warriors under the command of 10th Special Forces—with a Peshmerga force of some 65,000 fighters supported by the 173rd Airborne, 26th Marine Expeditionary Unit, a British special operations contingent, the 352nd Special Operations Group, the 3rd Battalion of 3rd Special Force, and the 2nd of the 14th Infantry (Golden Dragons).

highly unusual and controversial, but I would learn later that the invasion's commander (Lieutenant General David McKiernan) believed that 10th Special Forces was best positioned to use the 173rd Airborne effectively in the north.[21]

Task Force Viking entered Iraq in force on March 20, 2003, with the main body of 10th Special Forces forced to infiltrate via Jordan after Turkey denied overflight permission on three successive nights after the invasion had begun in the south and west.[22] Ours would be a nighttime, three-hour, low-level infiltration over robust Iraqi air defenses and a ride I will never forget.[23] Appropriately and famously dubbed the Ugly Baby, the operation demonstrated the potency of the partnership that we developed with the pilots of the Air Force's 352nd Special Operations Group during the previous year's joint unconventional warfare exercise. Its commander, Colonel O. G. Mannon, had resisted Air Force pressure to organize along doctrinal lines and instead rendered a brilliant dual-hatted performance as the deputy commander of Task Force Viking and commander of the special operations air component.[24] About 18 months earlier, 5th Special Forces had pioneered a similar arrangement in its infiltration into Afghanistan. The display of courage and audacity resulting from Ugly Baby forced the Turks' hand to allow for overflight to support our operations. They granted overflight permission for combat aircraft within a week, giving us access to

---

[21] His confidence in Task Force Viking and SOCCENT's other special operations task forces in what was to become Operation Iraqi Freedom earned him the eternal respect of the Special Forces leaders who worked for him.

[22] Although they kept denying permission both for the invasion and in the days that followed, the Turks promptly approved overflight when Ugly Baby forced their hand. One of our six aircraft was damaged by antiaircraft fire to the extent it needed to divert to the U.S. air base in Incirlik, Turkey. Perhaps the Turks did not want the specter of a plane full of American soldiers being shot down because of their intransigence, but our overflight was not an issue after Ugly Baby.

[23] My guys would later remind me that, being on the lead aircraft, I basically woke up the enemy in time for them to get shot. I was on the only bird in the flight of six that was not damaged.

[24] Doctrinally, as the commander of the air component of the Joint Special Operations Task Force, he would take his orders from the SOCCENT Air Force component in Qatar.

critical air support off of the aircraft carriers USS *Harry S. Truman* and USS *Theodore Roosevelt.*

The ground fight in northern Iraq, in contrast, went largely as planned, validating Task Force Viking's irregular warfare campaign approach. The first objective was the destruction of an enclave of Ansar al Islam, an al Qaeda affiliate and long-time enemy of the Kurds in the mountains that bordered Iran.[25] Following a number of Tomahawk strikes targeting their camps, a single Special Forces company and 6,500 Peshmerga, supported by U.S. gunships, devastated the organization, and the Ansar al Islam affiliates who survived by fleeing into Iran would sit out the rest of the war.[26] The tactics and organization for the ground operations were pure Peshmerga and decidedly different from those of the U.S. military. Our central challenge was to fold what support we could offer—primarily, intelligence, close air and sniper support, and tactical advice—into their formations. The combination of the warrior spirit and iron will of our Peshmerga allies and the masterful work of their Special Forces advisers would prove an early indicator of the significant power of a relationship that started in 1991 and continues to this day.

Task Force Viking's irregular warfare approach was similarly effective in routing Iraq's 13 northern divisions, as packets of Special Forces and Peshmerga fighters directed relentless air strikes from the two aircraft carriers in the Eastern Mediterranean. A key concession that I had made with our Turkish counterpart during our negotiations, a Special Forces colonel who had long trained with American

---

[25] Jonathan Schanzer, "Ansar Al-Islam: Back in Iraq," *Middle East Quarterly*, Winter 2004.

[26] Ansar al Islam camps had been struck by Tomahawks on the first day of the invasion, and before Special Forces teams were in place to prevent their escape (Timothy D. Brown, *Unconventional Warfare as a Strategic Force Multiplier: Task Force Viking in Northern Iraq, 2003*, MacDill Air Force Base, Fla.: Joint Special Operations University Press, September 2017). We knew at the time, from our intelligence preparation, that Abu Musab al Zarqawi and other al Qaeda operatives were being protected by this group. And although the Tomahawk missile strike was devastating, it allowed him and others to escape across the Iranian border—they were based about 500 meters from the border, and we did not have the helicopter assault capability to seal the back door.

Special Forces,[27] is that we would have no more than 150 Peshmerga per team. We kept to the spirit of the agreement, as we wanted to make sure that we maintained our part of the bargain, if nothing else than for larger political reasons with our contentious NATO ally. That said, the lads became very liberal on what constituted a Special Forces team, with two-man "teams" becoming increasingly common as the fight progressed.

Mosul and Kirkuk were liberated on April 10, the day after the fall of Baghdad and just three weeks after we had infiltrated, and my two 10th Special Forces battalions became responsible for securing them, each implementing a plan that they had formulated in advance of the invasion in consultation with the Kurds and CIA. Liberated terrain would be turned over, where possible, to tribes and ad hoc collections of citizen leaders, with civil affairs teams working with non-governmental organizations to hasten the return of some form of local governance. This push to restore local governance was part of our overall stabilization effort, which would remain a continual process until we turned over Mosul to then–Major General David Petraeus and the 101st Airborne Division and Kirkuk to then–Major General Raymond Odierno and the 4th Infantry Division.

In Mosul, we entered the city with a force of 30 Special Forces soldiers and about 600 Peshmerga fighters. But we learned quickly that our force of 600 Peshmerga fighters would not be enough. The Arabs had apparently not gotten the memo that "Mosul was liberated," and we were in a firefight at the municipal center within two minutes of my call to the Special Operations Command Central (SOCCENT) commander to tell him of our success. To their everlasting credit, the men of 3rd Battalion of 3rd Special Forces (attached to Task Force Viking for the invasion) did an overnight move from Kirkuk to Mosul to reinforce our lightly armed Special Forces and Peshmerga force. By the next morning, they were patrolling Mosul with armored vehicles.[28]

---

[27] The Turkish government would jail him a few years later for alleged participation in a coup attempt.

[28] The 3rd Special Forces battalions had up-armored Humvees and support vehicles that allowed effective mounted operations, which proved helpful in securing Mosul. I had ini-

The 26th Marine Expeditionary Unit arrived a few days later out of the Eastern Mediterranean and did everything that we asked—and did it well, including reoccupying the now psychologically critical municipal building.[29] We would have several firefights around that building over the next week, and we would stay busy until we were relieved by the 101st Airborne two weeks later.[30]

Kirkuk, we knew, could not be held with Kurdish forces. Our Kurdish friends, though extremely politically astute and perhaps the world's most resilient survivors, had a deep-seated and increasingly restless desire for independence and felt that they had claim to the city and its oil.[31] This was a political problem that we wanted to avoid,[32] and the arrival of the 173rd Airborne Brigade gave Task Force Viking the combat power needed to secure Kirkuk.

The 173rd Airborne Brigade jumped into Bashur Airfield to reinforce us six days after our arrival. The airfield, which was our main lifeline for resupply out of Europe, had already been secured by a Special Forces company and a Kurdish partner force of about 2,500. But the brigade's arrival was nonetheless impressive. We could hear their C-17s and then the soft pop of the chutes opening but could see only the occasional passing red or green glow from the onboard jump lights. It was the largest airborne assault since World War II, and I was told that it was to send a strategic message. It was a wet and chilly night and, after leaving the warmth of one of the Kurd's chai tents to greet our arriving compatriots, we found the 173rd commander huddled

---

tially assigned this battalion to secure the oil fields near Kirkuk, and they did an overnight refuel in Erbil to ensure that they arrived in time to support the following morning.

[29] The 26th Marine Expeditionary Unit was part of the theater reserve.

[30] Mary Beth Sheridan, "For Help in Rebuilding Mosul, U.S. Turns to Its Former Foes," *Washington Post*, April 25, 2003.

[31] Turkey, in particular, was concerned that this could embolden Kurds seeking independence from Iraq and "promoting similar ambitions among its own Kurdish population" (CNN, "U.S. Reinforcements Arrive in Kirkuk," April 10, 2003).

[32] It took some doing and explaining that the Kurds would be better off letting diplomacy and negotiation settle Kirkuk's fate, and I naively told them it would be settled by the ballot box. History would again prove the adage that the only real friends the Kurds have are their mountains.

under a poncho with a few of his officers, his red-filter flashlight, and a map doing what airborne commanders have done since Normandy.

The support of the 173rd Airborne Brigade proved important to the success of Task Force Viking's campaign, as it eventually assumed responsibility for the city of Kirkuk and strategic Kirkuk oil fields after they were seized by our Special Forces soldiers and Peshmerga allies.[33] It was here that I appreciated the support of the notoriously tough Special Forces Brigadier General James (Jim) Parker, the deputy commander of SOCCENT at the time and the airborne commander for the 173rd Airborne Brigade's jump. He had joined me in greeting the leadership of the 173rd Airborne on the drop zone to ensure smooth relations in what was viewed by many within the Army as a highly controversial decision to place its premier airborne brigade under the tactical command of a Special Forces group.[34]

I believe that the 173rd Airborne quickly learned the benefits of having a competent indigenous partner. Indeed, the brigade had to rely largely on indigenous transportation (coordinated by our Special Forces and their Peshmerga) to move from its drop zone on the airfield in Kurdish held territory behind the Green Line to Kirkuk.[35] Our admittedly uneasy but working relationship lasted until the 4th Infantry Division, after making its way north from Kuwait, took operational control of the 173rd Airborne Brigade.

Before redeploying to Kuwait, Brigadier General Parker also helped us defuse a situation that I believe could have upended our

---

[33] Kirkuk, we knew, could not be held by our Kurdish partners, given their deep-seated desires for independence and claims on Kirkuk.

[34] Brigadier General Parker was a former enlisted man, Vietnam veteran, and notoriously tough Special Forces leader, and the Special Operations Command Central deputy commander. He had been assigned by the SOCCENT commander to be the airborne commander for the brigade's jump, to ensure smooth relations in what was viewed within the Army as General McKiernan's highly controversial decision to place its premier airborne brigade under the operational control of a joint special operations headquarters and under the tactical command of a Special Forces group.

[35] Members of the 173rd were now foot infantry, and their assembly area at Bashur was about 30 kilometers behind the Green Line, the initial forward line of Peshmerga troops and their Special Forces advisers.

campaign in northern Iraq. Ours was an irregular warfare campaign, but it became evident that there were those in U.S. Army Europe who had a different approach in mind, as they started sending us heavy armor (e.g., M1 Abrams tanks, Bradley Fighting Vehicles) before we had finished bringing in the combat vehicles for our Special Forces battalions.[36] We had only a single airstrip in this area, the one we infiltrated into, and my Air Force component informed me that it had a very limited shelf life (in terms of the number of C-17 landings it could take). The diversion to heavy armor and their significant maintenance tail that was not essential to our immediate fight, at least in my view as the overall commander on the ground, were very nearly disastrous. Parker was able to engage with CENTCOM via his own headquarters (SOCCENT) to force a halt on heavy armor in favor of our vehicles—this is something that we could never have done on our own in time.

Task Force Viking also managed to neutralize Mujahedin-e Khalq, another terrorist group with a significant footprint in the north. It was an Iranian terrorist group laagered in our area of responsibility that Saddam had used as a tool against his Iranian neighbors, and its members had approached us and asked for a cease-fire. My thinking was that they might have some utility for us, as the long-term enemy in the region was Iran, and destroying the "enemy of my enemy" made little sense.[37] Lieutenant Colonel Ken Tovo, the commander of 3rd Battalion, masterfully orchestrated the cease-fire arrangements and then soon turned the problem over to Major General Odierno and the then-arriving 4th Infantry Division.

---

[36] U.S. European Command owned the departure airfield and was obviously responding to a different chain of command. Once started, the flow was very difficult to turn off.

[37] At one point, we had the Iranian-supported Badr Organization, likely with one of the Kurdish political parties turning a blind eye to its movement over the border, lined up to take on the Saddam-supported Mujahedin-e Khalq. I determined it was not in our interest to allow the Badr Organization to have any part of our success and, at a meeting with their leader, gave them an ultimatum to leave or risk being targeted. They left.

Inadequate Irregular Warfare Capabilities Above the Tactical Level: Difficulties in Achieving Enduring Successes

We left Iraq on May 23, arriving in Colorado Springs on the same day, and a week later we were toasting (from a ski lodge at Arapahoe Basin) our Special Forces mates in contact half a world away. I had made good on my promise to most of my troops that we would return home before the last ski run at Arapahoe Basin,[38] and together we celebrated the end of 10th Special Forces' role in the war in Iraq. Little did we know that the group would return to Iraq eight months later (after I had left command) to relieve 5th Special Forces, as it would every year until 2011.

In truth, the irregular warfare campaign was just beginning. But this would be of our making, as the United States would find itself on the defensive and grossly unprepared for the form of warfare that it would face.

The difficulties that were soon to come were foreshadowed during a breakfast hosted by CENTCOM's deputy commander shortly before I left. During the breakfast, also attended by the SOCCENT commander and the commanders of his other two special operations task forces, I suggested that security would quickly improve once we reconstituted the Iraqi Army. He paused, and in one of those eerily foreboding matter-of-fact ways, replied: "Charlie, what makes you think we are going to reconstitute the Iraqi Army?" In Mosul, we had put out a TV ad asking for Iraqi Army officers to report to the airport for instruction, and some 3,000 showed up, dutifully listened, and then were sent away to await further instructions. Instead of leveraging this indigenous force, we would let the Iraqi Army melt away into the population—waiting, watching, harboring raw hatred from the recent carnage visited on them, and leaving many to turn to the insurgency if only to put food on the table. It was evident that the United States was sowing the seeds of insurgency.

---

[38] When it became clear that we were leaving, one of our warrants reached out to the Colorado ski lodge and told them of the commander's promise. They were happy to offer us a good deal on ski passes.

I have to admit that I poorly prepared my staff to effectively coordinate our international partners in this mission. A UK special operations unit, for which we were notionally given tactical control, was memorably problematic. Its members seemed intent on being involved in the inevitable arbitration of oil resources, attempting to move twice toward the Baiji oil refinery quicker than the enemy situation allowed. Rather than staying with the tribes that U.S. intelligence had identified for them as partners, they twice prematurely pulled away from their tribal hosts and had to be pulled out in both cases because of enemy activity. The hallmark of Task Force Viking operations was the amount of autonomy necessarily given subordinate commanders to exploit opportunity, so long as it was within the commander's intent. The Brits, however, for whatever reason appeared to be operating on Home Office intent.

Historians will not be kind to those who thought that invading Iraq and removing Saddam Hussein was a good strategic move for the United States. The claim of weapons of mass destruction proved too thin: Some old chemical rounds were found but nothing to the scale presented to make the case. And the destabilization caused by our invasion was a direct result of poor policy decisions, foremost among them the failure to immediately reconstitute the Iraqi Army and use existing Iraqi expertise, primarily Baathist party members, to help govern.

CHAPTER NINE

Special Operations Campaigning in Latin America

In 2005, I was frocked to brigadier general and headed to Homestead, Florida, to take command of Special Operations Command South (SOCSOUTH).[1] I would now be responsible for most of America's special operations, and in my mind all of America's irregular warfare campaigns, in the Caribbean, Central America, and South America, building indigenous military and paramilitary law enforcement networks that could further mutual interests.[2] SOCSOUTH already had a rich history of conducting irregular warfare,[3] and U.S. special operations elements would operate in all but three countries of the region during my three years in command.[4]

By the time I took command, U.S. Southern Command (SOUTHCOM) had become an economy-of-force theater for the

---

[1] My immediate predecessor, Brigadier General Sal Cambria, had been moved to the SOUTHCOM J-3 (Operations) earlier than planned when a vacancy emerged. I had only a week for the frocking ceremony, to pack my bags, and to see my son graduate from medical school in Chicago before heading to Florida. It was wonderful getting a star, but hearing my son pronounced "Dr. Jeremiah Thomas Cleveland" was by far my proudest moment.

[2] These were referred to as either *theater* special operations or *continuous* special operations missions (Joint Publication 3-05, *Special Operations*, Washington, D.C.: Joint Chiefs of Staff, 2003, p. III-4). These missions, which required a deep understanding of the history, culture, and political realities of each country, demanded a good measure of continuity of thought, practice, patience, and persistence to shape each relationship based on today's interests and the reality that unknowable future security challenges may require allies in unlikely places.

[3] SOCSOUTH had conducted irregular warfare missions in every country in the region at one time or another.

[4] The three exceptions were Cuba, French Guiana, and Venezuela.

Department of Defense and for the U.S. special operations community in particular. This was a sharp contrast from the previous four decades: counterinsurgency efforts in Latin America had been a U.S. priority since the early 1960s, and counternarcotics had emerged as a national priority in the mid-1980s. U.S. special operations played a prominent role in both.[5] However, despite our history in the region, it was widely understood that the last thing the United States wanted was another draw on resources outside the "main effort" in the Middle East.[6]

But as fortune would have it, a high-profile, multiyear irregular warfare campaign in the jungles of Colombia would dominate my tenure at SOCSOUTH. Two years prior, in 2003, a simmering conflict in those jungles was pushed beyond its domestic bounds when the Fuerzas Armadas Revolucionarias de Colombia (FARC) downed a Department of Defense–contracted aircraft, executed an injured crewman, and took captive the three other Americans. The United States obtained its first credible intelligence on the kidnappers and their likely location shortly after I assumed command, marking the beginning of a three-year, SOCSOUTH-led irregular warfare campaign to recover our personnel.[7]

---

[5] The earliest significant commitment of U.S. special operations to Latin America was with the establishment of the Special Action Force, which was responsible for training Latin American forces on counterinsurgency. The Army established an additional Special Forces group, 8th Special Forces, for this mission. This unit was colocated with the School of the Americas in Fort Gluck, Panama (Ian Bradley Lyles, *Demystifying Counterinsurgency: U.S. Army Internal Security Training and South American Responses in the 1960s*, Austin: University of Texas, Austin, 2016).

[6] In truth, there were not any serious tasks being demanded of SOCSOUTH, leading me to conclude during the early days of my assignment that I could be a successful SOCSOUTH commander by doing the minimum necessary to keep the lights on while staying out of the papers. This did not sit well with me, nor with my staff.

[7] SOCSOUTH was assigned as the "lead for all [Department of Defense] planning and hostage recovery efforts" for the operation (Hy Rothstein and Barton Whaley, *The Art and Science of Military Deception*, Norwood, Mass.: Artech House, 2013, p. 386). Initially, in 2003, about one-third of the SOCSOUTH staff was involved in recovery efforts. However, overshadowed by problems half a world away, interest and emphasis waned, which disgusted some of SOCSOUTH's key staff—foremost among these was a Navy SEAL lieutenant commander, who would later play a key role in our three-year campaign to recover our people.

Operation Willing Spirit, as this rescue operation would be called, would involve more than 1,000 U.S. personnel at the operation's high-water mark.[8] However, it would be SOCSOUTH's Colombian partners, with limited U.S. help (though enabled by our shared three-year campaign), who would liberate the three Americans and 11 other hostages in a unilateral Colombian operation in July 2008. Less than three weeks after the daring rescue, the three former hostages attended my change of command at SOCSOUTH headquarters in Homestead, Florida. Our command and the many units and agencies that supported us in our campaign to get them back could have received no greater tribute.

Our success in Colombia demonstrated the critical value of a special operations headquarters whose mission was to provide indigenous-centric options in circumstances below the threshold of war. SOCSOUTH was able to facilitate the eventual indigenous option that rescued the hostages, as the command both owned the relationships with the host-government militaries and was able to synchronize U.S. capabilities (e.g., intelligence, logistics) with that of our partner forces and create decision space and options for policymakers.

My time at SOCSOUTH also validated, at least to me, the value of special operations campaign plans. In addition to communicating how our irregular warfare capabilities could create options for the SOUTHCOM commander, my campaign experience in northern Iraq and my recent scars fighting resourcing battles as the chief of staff of U.S. Army Special Operations Command (USASOC) had taught me that such a plan was an imperative in the inevitable fight for resources. Ultimately, I was proven right, as the SOUTHCOM commander's support for our plans proved essential in securing the resources that we needed from, first, SOCOM and, later, the Department of Defense and Department of State.

My time at SOCSOUTH also showcased the critical role of relationships in our ability to provide indigenous-centric options. Part-

---

[8] Notably, this deployment, which occurred during January to March 2008, did not make the papers in the United States or Colombia, a true testament to the professionalism of our Colombian partners and the elite American troops involved.

nerships with our allies' militaries and local communities in those countries were already the bread and butter of SOCSOUTH, and we maintained relationships first developed in the 1960s and carried through the 1970s and 1980s. But gaining the support of the U.S. ambassadors in the countries that I operated, the Department of State's regional bureau, the SOUTHCOM commander, and the commander of SOCOM proved critical. Equally as important were partnerships with peer organizations, both military (e.g., other SOUTHCOM component commanders) and civilian (e.g., DEA, CIA). Each relationship required attention and would rapidly atrophy without senior-leader involvement, but each was critical to our efforts in its own way.

Success Through Partnering: Operation Jaque

Operation Jaque (or Operation Checkmate, in English), which liberated three Americans and 11 other hostages from the FARC in July 2008, was a Colombian operation. However, this success had its foundation in a deep partnership between the special operations forces of our two countries, having spent years operating together in the hunt for Pablo Escobar and fighting both drug cartels and domestic insurgencies. It was this partnership, which built mutual trust and the capabilities of the Colombian elements that would execute the operation, and the support of the U.S. Embassy that would ultimately allow an audacious Colombian plan to succeed.

In 2003, immediately after the three Americans were seized, the SOCSOUTH commander at the time, Brigadier General Remo Butler,[9] flew to Bogotá to assess options and take charge of the U.S. military portion of any recovery effort. However, a lack of available U.S. intel-

---

[9] Brigadier General Butler was one of the most colorful characters with whom I served and one who had been very helpful to me when I was a newly promoted field grade. We served together in Panama, where he was the SOCSOUTH operations officer (J-3), while I was the operations officer for 7th Special Forces, 3rd Battalion, and he led our planning team's visit to Fort Bragg when we briefed Task Force Black's plan for Operation Just Cause. Ambassador Anne Patterson, the chief of mission when the hostages were captured, and the Colombians had tremendous respect for Remo. I tried to build on that as best I could.

ligence and special operations assets, and the very tough jungle terrain in which our comrades were captured, stymied any immediate efforts, and the window for a quick recovery rapidly closed.

The first opportunity to recover the hostages arrived in 2005, shortly after I had assumed command, when an interagency intelligence fusion cell at the U.S. Embassy determined that it had a source who could locate the hostages.[10] It just so happened that SOCSOUTH planners were training with the Colombian national special operations headquarters on intelligence preparation, operations planning, and synchronization, and we immediately pivoted to determining military options should the hostages be located. This lead, in the end, did not pan out, but it demonstrated the potency of SOCSOUTH's working relationship with the Colombians and how it could be rapidly leveraged to develop an irregular warfare option based on U.S.-developed intelligence.

In June 2006, another opportunity to rescue the hostages presented itself when U.S. intelligence identified a handoff point for the transfer of the hostages.[11] With specific intelligence on the location and a two-day window in which the handoff was expected, our recommendation was to support a Colombian special operation to possibly snatch our hostages. To share in the risky "go or no-go" decision and to most effectively apply U.S. support to any rescue attempt, SOCSOUTH recommended that six 7th Special Forces soldiers accompany a 200-person Colombian rescue force. The combined force mission was approved by the U.S. ambassador, William Wood, and, on his strong recommendation, by the SOUTHCOM commander.[12]

---

[10] This was the U.S. Embassy Intelligence Fusion Cell, which provided an early model for interagency intelligence coordination that would be later popularized in Iraq and Afghanistan. This fusion cell provided an "intelligence training and tracking network that allowed Colombia to push tactical intelligence to local commanders in real time, utilizing signals-intelligence intercepts" (Shawn Snow, "A Plan Colombia for Afghanistan," *Foreign Policy*, February 3, 2016). A more detailed description of this capability is provided in Dana Priest, "Covert Action in Colombia," *Washington Post*, December 21, 2013.

[11] This was U.S. intelligence but greatly aided by the Colombian leadership's in-depth understanding of its enemy.

[12] I was told that the SOUTHCOM commander's decision was vetted in Washington, D.C.

This mission would come up empty,[13] but it demonstrated the true potency of SOCSOUTH's partnerships. Deploying our forces, the six soldiers from 7th Special Forces Group, required approval at the highest levels of the U.S. government, as it put U.S. troops on the ground in hostile territory outside a declared war zone. This simply would not have been possible without the strong support of Ambassador Wood. The experience would also lead to the establishment of a (in my experience) unique combined U.S.-Colombian special reconnaissance team of three U.S. Special Forces soldiers and three Colombian special operators.[14] This cooperation would become a centerpiece in the success that was to come.

The next opportunity emerged in January 2008, almost a year and a half later, in the wake of significant pressure by the Colombian Army against the insurgent element holding the bulk of the high-profile U.S. and international hostages. The U.S. and Colombian intelligence communities learned that the hostages would be moved as a result of this pressure, and a mix of combined U.S.-Colombian special operations teams were tasked and deployed with Colombian special operations teams to interdict likely travel routes. In all, 33 Americans would be in the field on missions ranging from a week to more than a month.[15]

This mission was partially successful. One of the Colombian reconnaissance teams observed the American hostages along the Apaporis River, providing the first confirmable proof of life since their capture five years earlier. For four days, the hostages were observed

---

[13] The force infiltrated, moved to the handover spot, and set up for an ambush. The team observed a 20-person FARC element moving through the area that later was assessed to have been the FARC rear security detachment. The force exfiltrated without making contact. The U.S. side learned some hard lessons about operating in the jungle.

[14] I would visit the team in 2007 and can honestly say that it was impossible to pick out the U.S. operator from his Colombian counterpart.

[15] The latter was a three-person team from 19th Special Forces, and which spent more time in hostile territory than any mission since Vietnam. Even in Iraq and Afghanistan, missions did not stay more than 30 days outside the wire. These three soldiers had lost an average of more than 30 pounds and would contribute more than 25 pages of single-spaced lessons learned. We had a new appreciation for our Colombian counterparts.

by our Colombian partners.[16] Unfortunately, the time at that location proved too short, distances too great, and the number of unknowns too significant to risk a jungle rescue. However, this mission set the conditions for Operation Jaque just a few months later, as it facilitated a raid by the Colombian Army that convinced the insurgents that the Americans and Colombians were in pursuit and that it was too risky to move the hostages from an area that was becoming increasingly well-known to our Colombian partners.

In July 2008, relying on deception with Colombian special operators and intelligence professionals posing as a humanitarian nongovernmental group, the Colombians liberated the three Americans and 11 other hostages. Promising to move the hostages to a safer location, this Colombian covert force picked up the hostages in a specially prepared helicopter, disarmed the captors, and flew the hostages to safety. This was an audacious plan made possible by years of work to provide decisionmakers options at the time of need and give them confidence in the mission's chances of success.

U.S. special operations would play a tactical role in this operation, but more important was the mutual trust necessary for this unorthodox and risky operation. Ambassadors Patterson, Wood, and Bill Brownfield; the intelligence professionals at the U.S. Embassy; and the military units training, advising, and learning alongside their Colombian counterparts all had a hand in setting conditions for this rescue option. SOCSOUTH's recommendation to Ambassador Brownfield was that the risk was worth taking, and he agreed.[17] I provided the same recommendation to a more skeptical Admiral James Stavridis, the SOUTHCOM commander, and his staff. He, too, would eventually agree, and the White House would eventually follow Ambassador Brownfield's recommendation to proceed.

---

[16] The hostages were first observed bathing under guard, conversing in English (Charles H. Briscoe and Daniel J. Kulich, "Operación Jaque: The Ultimate Deception," *Veritas*, Vol. 14, No. 3, 2018).

[17] Colonel Greg Wilson, who was my Special Operations Command Forward (SOC Forward) commander in Colombia at the time, was instrumental in both assessing details of the Colombian plan and gaining Ambassador Brownfield's support for it. This is discussed in more detail later in this chapter.

Creating Indigenous-Centric Options: Establishing a Special Operations Campaign Plan

During my three years in command, I came to believe that the primary role of SOCSOUTH, as well as the other theater special operations commands,[18] was to provide indigenous-centric options in support of the combatant commander's campaign objectives. SOCSOUTH's responsibility was to provide the four-star commander of SOUTHCOM in-house staff expertise on the best use of all U.S. special operations forces and a general officer–level special operations headquarters to command and control special operations during a regional crisis. But I quickly realized that SOUTHCOM's staff was much less comfortable with our irregular warfare capabilities, and SOCSOUTH's comparative advantage was in providing options that leveraged these capabilities.

To communicate how SOCSOUTH's irregular warfare capabilities could create options for the SOUTHCOM commander, we established campaign plans for our irregular warfare approaches in the Caribbean, Central America, and South America.[19] Our campaign plans would differ from standard plans of exercises and engagements that such components as SOCSOUTH would typically develop in that it would provide the *why*, detailing how our proposed engagements and indigenous capacity-building efforts would support SOUTHCOM's component of the national counterterrorism mission, the already-decades-old counterdrug support mission, and broader country security assistance objectives.

There were to be three SOCSOUTH campaign plans, each aligned with a specific region and the unique operational circumstances

---

[18] At the time, there were five active theater special operations commands: Special Operations Command Central (SOCCENT), Special Operations Command Europe, Special Operations Command Korea, Special Operations Command Pacific, and SOCSOUTH. Two additional theater special operations commands would later be established: Special Operations Command Africa in 2008 and Special Operations Command North in 2012.

[19] Our plan would be modeled after a similar plan developed for Operation Enduring Freedom–Philippines.

and U.S. objectives for that region.[20] The first was the Andean Ridge plan, which nested the hostage rescue efforts within the regional counterdrug mission and a broader effort to create policy options for influencing events in the increasingly problematic Venezuela, Bolivia, and Ecuador. The main effort was Operation Willing Spirit, in which our irregular warfare capability was effectively supported by our nation's world-class raiding capability.[21] But strengthening the special operations capabilities of partners along Venezuela's borders was also critical, including a focused engagement with Ecuadorean special operations.[22]

The second was the Southern Cone plan, focused on strengthening partnerships with the already capable special operations forces in Argentina, Brazil, Chile, Paraguay, and Uruguay. The centerpiece of this campaign plan was a multinational exercise called Southern Star, which was designed to build the capabilities and promote interoperability among the militaries of this region.[23] One of the key goals was

---

[20] These three campaigns were aligned with the overall command and control structure for SOCSOUTH (see, e.g., Christian M. Averett, Louis A. Cervantes, and Patrick M. O'Hara, "An Analysis of Special Operations Command—South's Distributive Command and Control Concept," thesis, Monterey, Calif.: Naval Postgraduate School, 2007).

[21] The unit's commander and a one-star from higher headquarters agreed to collaborate on this priority mission. We would invite them to meetings with our Colombian counterparts and embassy staff, serving as intermediaries with the host nation and facilitating access to local training facilities. That team would provide intelligence and logistics where it could. Its mission was to attempt a rescue if conditions allowed, but I—the SOCSOUTH commander—would remain in charge as long as the Colombians had the lead. In retrospect, this supported-supporting relationship worked well. I hope that I would have been quick to reverse the roles if an indigenous solution was not feasible. Others were critical of the weakness of our arrangement, but I highly respected the spirit of cooperation that emerged from our relationship.

[22] Working with the U.S. country team in Bolivia for noncombatant evacuation plans was also a component of this effort.

[23] Victoria Meyer, "Southern Star Shines Brightly in Chile," *Tip of the Spear*, November 2009. The exercise was initially hosted by the United States, but Chile would take the lead in 2009. Years later, shortly after I retired, a Chilean special operations officer sought me out to tell me that this exercise and SOCSOUTH's collaboration and training at the operational level had been the inspiration for the eventual development of a Chilean version of SOCOM (Alexis Ramos, "Joint, Combined Forces Conclude Southern Star," U.S. Southern Command, September 7, 2018).

to get the already-capable special operations professionals in Chile and Brazil, and later Argentina, to help advise other nations first in their region and then maybe around the world. This plan also included a targeted push to build a capable Paraguayan special operations interdiction capability that could be leveraged if required for operations on the Paraguayan portion of the Argentina-Brazil-Paraguay triborder area, a region with known connections to Lebanese Hezbollah. It was to be the best example of how the then newly available Section 1208 authority could be used in a relatively short period to build a well-resourced and well-trained interdiction capability.[24]

The third campaign is one that we billed as being SOUTHCOM's contribution to Operation Enduring Freedom. It focused on improving the "capabilities of Caribbean and Central American partners to interdict and disrupt terrorists who might leverage illicit transnational routes and uncontrolled areas to threaten the United States,"[25] although drug traffickers, human smugglers, and gun runners were very much part of our mandate. Operating out of Honduras's Soto Cano Air Base, this campaign was focused on building a "picket fence" along the southern approach to the United States by renewing relationships with host-nation special operations capabilities and developing partnerships with U.S. embassies across the region. Partnering and supporting host-nation counterparts during real-world interdiction missions were critical to our approach. The best and most controversial example was the U.S. Special Forces training and advising of the Guatemalan Kaibil in support of combined Guatemalan-U.S. counternarcotics missions. An important principle behind this plan was creating options well before

---

[24] This authority comes from Pub. L. 108-375, National Defense Authorization Act of Fiscal Year 2005, October 28, 2004. Anthony F. Heisler, "By, with, and Through: The Theory and Practice of Special Operations Capacity-Building," thesis, Monterey, Calif.: Naval Postgraduate School, 2014.

[25] James G. Stavridis, "Statement of Admiral James G. Stavridis, United States Navy Commander, United States Southern Command," Washington, D.C.: U.S. House of Representatives, Committee on Appropriations Subcommittee on Defense, March 5, 2008.

the time of need, which required strategic patience and some vision on the part of those who allocate the funds.[26]

Briefing our campaign plans to the SOUTHCOM commander turned out to be a difficult enterprise. His principal staff believed, as was the doctrine of the time, that there could only be a single campaign plan for the theater and that it was SOUTHCOM's responsibility alone to design and execute this campaign. Doctrinally, they were of course right, but the reality was that a single campaign plan could not cover the complexities of the Caribbean and Central and South America. Further, the existing SOUTHCOM campaign plan did not make full use of SOCSOUTH's irregular warfare capabilities in engaging the insurgencies, illegal drug trafficking, and resistance movements across the region. In this sense, these plans were necessary to complete any viable SOUTHCOM plan.

A window opened in late 2005 to brief our three campaign plans to the SOUTHCOM commander. He had directed all subordinates to brief him on their theater engagement plans—the planned exercises and engagement—for the next fiscal year. We briefed our campaign plans instead. The meeting lasted nearly two hours and quickly devolved into a discussion between me and General Bantz Craddock, with his entire primary staff in attendance. He asked hard and, where he had interest, pointed and detailed questions. A friend, General Craddock's two-star director of operations and my predecessor at SOCSOUTH, would tell me afterward that the meeting to him was one of the most painful he had attended with his commander.

At the end of the session, General Craddock stood up, smiled, and told his staff that he liked it and said that this is what we are going to do. That success would provide a blueprint for my next tour, at Special Operations Command Central (SOCCENT), and be a key driver behind the institutional changes we would make during my command of USASOC years later.

---

[26] Such missions routinely run into problems when times get lean and are made tougher when trying to describe the military mission and finding there to be no real label in U.S. operational concepts for it.

Resourcing Irregular Warfare: The Derivative Benefit of the Special Operations Campaign

When I arrived at SOCSOUTH, resources for the irregular warfare component of our mission (in terms of personnel and dollars) were extremely limited.[27] However, this was a challenge I understand only too well, as I had spent the previous two years (2003–2005) fighting for resources for USASOC's commando and irregular warfare units as its chief of staff. Sustaining a world-class counterterrorism capability with the global reach demanded by policymakers in the wake of September 11 was expensive and increasingly coming at the cost of the irregular missions, such as those we faced in Latin America. Our campaign plans would prove a critical weapon in gaining the resources from the Department of Defense and Department of State that SOCSOUTH needed to be effective.

My two years as the chief of staff at USASOC had been dominated by the dreaded Program Objective Memorandum. This is the process through which the Department of Defense, SOCOM, and the Army allocate resources, racking and stacking requirements against the finite resources available. My mission-essential task as the chief of staff was to develop a strategy that ensured that each of USASOC's nine "tribes" had the resources they needed,[28] and then convince SOCOM

---

[27] When I took command of SOCSOUTH, it was by far the most thinly resourced of the four theater special operations commands. Funding for the then-active four theater special operations commands (SOCCENT, Special Operations Command Europe, Special Operations Command Pacific, and SOCSOUTH) came from both SOCOM and their geographic combatant command (e.g., SOUTHCOM), with SOCOM responsible for special operations–specific requirements and the geographic combatant command responsible for the rest (Elvira N. Loredo, John E. Peters, Karlyn D. Stanley, Matthew E. Boyer, William Welser IV, and Thomas S. Szayna, *Authorities and Options for Funding USSOCOM Operations*, Santa Monica, Calif.: RAND Corporation, RR-360-SOCOM, 2014, p. 6). Securing funds from the geographic combatant command was a significant challenge and would remain so for at least another decade (Loredo et al., 2014, p. 7), although funds from SOCOM were also difficult to obtain, given that most resources were going to the main effort in Afghanistan and Iraq.

[28] USASOC's nine tribes at the time consisted of 112th Special Operations Signal Battalion, 160th Special Operations Aviation Regiment, 528th Sustainment Brigade, civil affairs,

to provide us the resources necessary to implement that strategy.[29] I intended to make this process as transparent as possible, so I brought all nine of the tribes to the same table in hopes of building a mutual understanding of each other's roles and needs. The weekly meetings were frequently contentious—a food fight that often seemed every bit as nasty as today's political campaigning.[30]

Achieving the right balance between the Army's irregular warfare capability and the Army's portion of the bill for America's national counterterrorism apparatus proved the most difficult component of this task. It was tough and, in my view, highly subjective work, which was complicated by SOCOM's renewed prioritization of our counterterrorism-focused capabilities in the then–global war on terror.[31] The fact that the mission and level of resourcing were classified

---

psychological operations, Rangers, Special Forces, and two special mission units. This was simplified in 2014, with the establishment of 1st Special Forces Command (discussed later).

[29] A negative unintended consequence of the Nunn-Cohen Amendment to Pub. L. 99-661, National Defense Authorization Act for Fiscal Year 1987, November 14, 1986, which gave SOCOM "service-like responsibilities to organize, train, and equip [special operations forces] worldwide," was that the services became much more reticent about funding special operations requirements, as they expected that these "requirements [would] be filled exclusively from the MFP-11 funding pool" (William R. Lane, *Resourcing for Special Operations Forces [SOF] Should Responsibilities Be Passed from USSOCOM Back to the Services*, Carlisle, Pa.: U.S. Army War College, 2006, pp. 1 and 6; William G. Boykin, *Special Operations and Low-Intensity Conflict Legislation: Why Was It Passed and Have the Voids Been Filled?* Carlisle, Pa.: U.S. Army War College, 1991, pp. 52–52).

[30] The hope was that this transparency would allow an apples-to-apples discussion of funding requirements, and then I could use logic, passion, and shaming to kluge together a strategy for USASOC. With different and often warring tribes competing for resources from the same pool, there was a real risk that this could be less than a collegial affair. But each week, the tribes would sit around the same conference room, and gradually they gained an appreciation of the arguments of their USASOC teammates. We held the meetings every Friday at 2 p.m. to help hone the arguments, and it was often a raucous affair that stretched well into the evening.

[31] In 2002, SOCOM was given the "lead in planning the war on terror" by Secretary of Defense Donald Rumsfeld (Bryan D. Brown, "U.S. Special Operations Command Meeting the Challenges of the 21st Century," *Joint Forces Quarterly*, Vol. 40, No. 1, 2006). SOCOM was established initially to be a force provider, and this shift toward being a warfighting command created a SOCOM-internal competition for resources and staff (see, e.g., Lane, 2006).

made frank conversations difficult and prevented the apples-to-apples conversations that I had hoped for. Further, despite being notionally in charge of the process, my influence over the staffing process of these classified units was often challenged.[32] Senior leaders at SOCOM would often intervene, exercising "senior military judgment" to support an ever-expanding constellation counterterrorism of activities, units, and staffs.[33] Yet few had the expertise to understand the consequences of the corresponding underinvestment in America's irregular warfare capability.

At SOCSOUTH, our campaign plans set out an ambitious agenda for our irregular warfare efforts in the theater, and the number of special operations teams and operational funds available to SOCSOUTH in the post–September 11 period would be insufficient for us to be successful. Although these plans allowed us to explain our requirements, it was General Craddock's approval for our approach that allowed us to augment SOCSOUTH's limited funding by securing additional resources at SOCOM's Global Synchronization Conference and later the Department of Defense and Department of State.

Even with this support, we still faced shortfalls in the number of teams and money we needed to fulfill the requirements of our approved campaign plans. Other theaters rightly had overall priority, although it seemed to me that the priority for our irregular capabilities should have been in this (and other) economy-of-force theaters where those capabilities would have had higher return on investment. The effect was that we had to split teams to satisfy some elements of the campaign, and other elements would remain undone. In Central America, for our contribution to Operation Enduring Freedom, we split teams by

---

[32] These units were run through essentially a shadow staff under the deputy commanding general, which were quick to call him if they felt that they were getting "meatloaf" rather than the "New York strips" that they were used to.

[33] During SOCOM's own resourcing process, senior leaders within SOCOM would similarly intervene to ensure that these capabilities were resourced. SOCOM relied on a computer-scoring system to rack and stack funding priorities, but they were adjusted by "senior-leader" judgment to make sure that the "cool guys" had everything they needed. The "Borg," as it was called by those of us outside its event horizon but clearly within its gravitational pull, had a seemingly insatiable appetite.

design to make sure that we could maintain a training and operational relationship with each of the region's special operations commands. In some countries, our efforts to build the counterterrorism capabilities of partners was stunted by our inability to access the limited Section 1208 funds. Even in Colombia, we struggled to obtain the augmentation necessary to staff our forward operational constructs, as so much of any extra staff officer capacity had to be committed to the new special operations headquarters support operations in Afghanistan and Iraq.

Relationship Building as a Mission-Essential Task: The First Special Operations Command Forward

Relationship building would be my personal mission-essential task as the SOCSOUTH commander, and these relationships would prove a powerful enabler for our efforts. Success in resourcing and executing our campaign plans did require support from the SOUTHCOM and SOCOM commanders, but success would also depend on support from their staffs; fellow component commanders of SOUTHCOM; the Army, Air Force, and Navy special operations commands subordinate to SOCSOUTH;[34] civilian leaders from across the interagency;[35] and the militaries, political leadership, and populations of our allies. Each relationship required attention, because each contributed to the SOCSOUTH mission.

During my tenure, we began experimenting with approaches that we might use for strengthening these campaign-essential relationships. The most successful was unquestionably the establishment of the first-ever Special Operations Command Forward (SOC Forward) at the U.S. Embassy in Colombia. U.S. special operations had a long-standing partnership with the embassy, as 7th Special Forces and a handful of talented Spanish-speaking SEALs played a central role in

---

[34] The Reserve Command at Homestead was also critical for our efforts.

[35] In this case, there were also civilian leaders from the Department of Defense, DEA, CIA, Customs and Border Patrol, and others assigned to the U.S. country teams across the embassies in SOUTHCOM.

Plan Colombia, a multiyear counterinsurgency program that was the U.S. Embassy's number-one priority,[36] and had an excellent relationship with their Colombian military and National Police counterparts. The SOC Forward element allowed us to deepen this partnership, building needed trust as a member of the embassy team, ultimately enhancing our ability to support the hostage mission.

The SOC Forward, in this instance, was a ten-person element, led by a colonel-level special operator from the SOCSOUTH staff, that worked directly for me but was instructed to take direction from the U.S. ambassador. This persistent presence of special operators and the relationship that they would develop with the ambassador, essentially becoming one of his assigned agencies and a part of his team, would prove critical for the successful rescue of the hostages three years later. Colonel Greg Wilson, my SOC Forward in Colombia at the time, provided the ambassador the expert military advice needed to make informed decisions, building support among the broader country team and facilitating Colombian efforts to develop this audacious rescue plan. In both cases that a rescue opportunity seemed possible, I augmented the SOC Forward with a small command and control element from SOCSOUTH to become my staff during the operation.

The success of the SOC Forward in Colombia is demonstrated most persuasively, perhaps, by the fact that it would be replicated in countries throughout the world. Similar elements were soon established at the U.S. embassies in Paraguay and Honduras to coordinate, respectively, SOCSOUTH's Southern Cone and Caribbean campaign plans. And during my next assignment at SOCCENT, SOC Forwards would be established in Jordan, Lebanon, Pakistan, Tajikistan, and Yemen.[37]

---

[36] Plan Colombia, which would run from 2000 to 2015, when it was transitioned to Peace Colombia, would be later heralded as one of the most successful U.S. counterinsurgency programs (David Sosa, "Peace Colombia: The Success of U.S. Foreign Assistance in South America," U.S. Global Leadership Coalition, May 10, 2017).

[37] Joshua Lehman, *Leading in the Gray Zone: Command and Control of Special Operations in Phases 0-1*, Naval Station Newport, R.I.: Naval War College, 2016, pp. 5–6.

However, this SOC Forward construct would not be without controversy. Critically, it added a third colonel to the country team, reducing the authority of both the U.S. military group commander and military attaché. It also ran counter to the intent of the Secretary of Defense, who had issued a directive in 2007 requiring that a single "Senior Defense Official" be identified to speak for the Department of Defense.[38] In later years, SOCOM's Special Operations Liaison Officer Program would complicate things further, by adding an additional colonel to the embassy staff.[39]

Our argument was that an ongoing special operations campaign required a forward special operations command element to coordinate the mission, that the senior defense official had neither the experience nor the resources to play this role, and that the theater special operations command (SOCSOUTH, in this case) was best positioned to provide a team with the requisite expertise. Our plan would also give the U.S. ambassador in that country a powerful tool—specifically, an experienced senior special operator and staff with the expertise to confidently make recommendations about indigenous or U.S.-supported special operations missions.[40] This approach was especially useful for irregular warfare campaigning, in which Special Forces and other irregular warfare capabilities were the primary players.

---

[38] Department of Defense Directive 5105.75, *Department of Defense Operations at U.S. Embassies*, Washington, D.C.: U.S. Department of Defense, December 21, 2007.

[39] The purpose of the special operations liaison officers, the first of which was assigned in 2007 (to the UK), was to build to create a liaison between SOCOM and the special operations headquarters of our partners across the globe (Paul J. Schmitt, *Special Operations Liaison Efforts: (SOLO) or Team Effort?* Carlisle, Pa.: U.S. Army War College, 2013, p. 10). Although these officers worked for the theater special operations command (Schmitt, 2013, p. 10), it would be my experience in later years that these officers eroded the influence of the theater special operations command at some embassies by offering a direct conduit back to SOCOM.

[40] My experience in both SOCCENT and SOCSOUTH would be that most of the colonels would make it work. When the relationship was not working, counseling the colonels worked reasonably well.

I had come to SOCSOUTH believing that we should be responsible for developing and executing options that leveraged indigenous mass, fires, intelligence, and logistics. The relationships that we developed proved critical in gaining the support that we needed from the Colombians, Department of Defense, and Department of State for a campaign plan that provided both an indigenous option (that would likely be timelier but with higher risk) and a unilateral operation by U.S. national assets. The success of Operation Jacque, arguably the most brilliant deception operation since World War II, in safely recovering our three Americans was a direct result of the trust built by these relationships. You had to be there to build them. I came away from three years at SOCSOUTH convinced in the power of leveraging indigenous or host-nation capacity, if done with patience and understanding. We had an American way of irregular war—the nation just needed to get serious about it.

CHAPTER TEN

At the Vanguard of American Irregular Warfare

In 2008, I received a second star and was ordered to take command of Special Operations Command Central (SOCCENT). This was the nation's main effort, and I was now responsible for overseeing American irregular warfare efforts in the two major theaters of war (Afghanistan and Iraq) and across the Arabian Peninsula, Central Asia, and Southwest Asia, which had become the primary battlegrounds for the counterterrorism aspects of Operation Enduring Freedom. I had very big shoes to fill, as my predecessor at SOCCENT was the legendary Lieutenant General John Mulholland, who had famously overthrown the Taliban in 2001 as the commander of 5th Special Forces Group.[1]

There were dramatic differences between SOCCENT and Special Operations Command South (SOCSOUTH), where I had just completed a three-year tour. For one, my staff at SOCCENT was much larger. I now had some 1,000 soldiers, sailors, airmen, marines, and civilians working directly for me, a nearly nine-fold increase from the 120 personnel assigned to me at SOCSOUTH, and a forward headquarters with a permanently assigned staff in Qatar. The operational tempo was also drastically different, as I now had operational requirements in every country in theater, some 20 in all, with almost 5,000 special operators in theater at any given time. In contrast, my forces at SOCSOUTH averaged around 200, unless an event related to Operation Willing Spirit was under way. There were also cultural differences, as our partners at SOCSOUTH did not hold your hand, kiss you on

---

[1] Mulholland, a long-time friend, had been promoted quickly up the ranks and was headed to the three-star job at USASOC.

the cheek, or hide their whiskey drinking, but although different in manner, the importance of befriending and developing trusted, competent indigenous partners remained paramount.

My time at SOCCENT again demonstrated the critical role of special operations headquarters in creating indigenous-centric options for U.S. policymakers, validating a key lesson from Operation Willing Spirit at SOCSOUTH. In priority countries when the Department of State had the lead (Lebanon, Pakistan, and Yemen), SOCCENT spearheaded designing and executing irregular warfare campaigns that aligned with the intent of both the U.S. ambassador and CENTCOM. For Afghanistan and Iraq, both declared theaters of war, we established new operational-level special operations headquarters to play this role as a consequence of the breadth and depth of U.S. operations in each country.

It was during my three years at SOCCENT, as I reflected on the stalemates in Afghanistan and Iraq, that I first realized a fundamental flaw in our approach to national security. We were failing in part because our approach did not properly account for the fact that these contests were about influencing and coercing people and not controlling territory. We had retained substantial thinking about how to operate in these conflicts, entombed in such seminal documents as the counterinsurgency field manual or *Small Wars Manual*,[2] and our tactical forces demonstrated remarkable capability and adaptability in both contests. But we did not have needed institutions at the operational level or above, within either the military or the broader national security apparatus, for this form of warfare. The result was that we struggled to develop and execute the type of irregular warfare campaign appropriate for these conflicts and largely failed to develop effective and appropriate policies. Our domain model was focused on dominating in physical domains—land, sea, air, and space—and we had not

---

[2]   Field Manual 3-24 and Marine Corps Warfighting Publication 3-33.5, *Counterinsurgency*, Washington, D.C.: U.S. Department of the Army, December 16, 2006; U.S. Marine Corps, *Small Wars Manual*, Washington, D.C., 2014.

developed the tools for success in this new "human domain" in which victory necessitated winning the "contest of wills."[3]

One other surprising lesson was, with the exception of the commander, how little advocacy we at SOCCENT would get from SOCOM for either our irregular warfare campaigning efforts or any efforts to improve broader U.S. capabilities in these fights—not opposition, just not any support. Some 20 years after the Senate Armed Services Committee had censured SOCOM for not giving sufficient priority to the "low intensity conflict" components of its portfolio,[4] many simply just did not see the development of America's irregular warfare capabilities as a priority. SOCOM's overwhelming focus on its manhunting capability was perhaps understandable, particularly given pressures from policymakers in Washington, D.C., in the post–September 11 era. But the result was that SOCOM, and consequently the United States, was not developing the institutional-level irregular warfare capabilities needed for success in these conflicts.

Irregular Warfare Campaigns in Lebanon, Pakistan, and Yemen: Adapting Lessons from SOCSOUTH

I came into command of SOCCENT with a clear understanding that our central value was in our ability to provide indigenous options to the CENTCOM commander.[5] The three CENTCOM commanders for whom I worked universally understood and tacitly approved of this role,[6] with each recognizing the value of an irregular warfare campaign that leveraged indigenous resources and approaches to achieve

---

[3]   Raymond T. Odierno, James F. Amos, and William H. McRaven, *Strategic Landpower: Winning the Clash of Wills*, Washington, D.C.: U.S. Army, U.S. Marine Corps, and U.S. Special Operations Command, May 2013.

[4]   Adams, 1998, pp. 205–206.

[5]   We would be responsible for maintaining the relationships in theater necessary to mitigate the "principal agent" problem endemic to controlling the actions of surrogates, which required a longer time horizon than the combatant command.

[6]   These three were General Marty Dempsey, General David Petraeus, and General James Mattis.

their objectives. The CENTCOM region also had a cadre of experienced U.S. ambassadors who well understood the role and value of SOCCENT for their own efforts.

We focused our efforts on three priority countries, at least initially, and began to develop irregular warfare campaign plans for Lebanon, Pakistan, and Yemen. We built on the success of the approach employed in Colombia with Operation Willing Spirit; during my tenure as the SOCSOUTH commander, the intent of these campaign plans was to provide the ambassador options. Each plan provided an option for accessing military resources without the arrival of a U.S. invasion force,[7] leveraging SOCCENT's relationships and resources in those nations and in neighboring countries, for problems that were of foremost importance to the ambassador.[8]

Central to these efforts was the establishment of a SOC Forward in each country, a construct that was critical to our success with Operation Willing Spirit.[9] The SOC Forward, which would embed in the country team, gave SOCCENT the ability to design and execute irregular warfare campaigns in each country, not as an outside agency but rather a teammate in the embassy's fight.[10] This element provided

---

[7]　Lehman, 2016, pp. 5–6.

[8]　This was described by some as a "second option." The first option, which was and is the focus of most Department of Defense planners, is a contingency operation relying solely or at least primarily on U.S. capabilities. We learned that no one else in the Department of Defense was identifying indigenous solutions—it was for us to do.

[9]　In Pakistan, we were able to retool the existing Special Operations Command and Control Element for this purpose. A Special Operation Command and Control Element is the doctrinal title for a Special Forces command and control node typically colocated with an Army corps that provides tactical control of Special Forces teams operating in that corps' area.

[10]　Around this time, versions of the SOC Forward that covered a group of countries in a given region began to pop up. This regional construct differed from the SOC Forwards in that the new versions would be unable to establish the needed close relationships with all the country teams in their region, as the leadership of these elements would be based in a single embassy and rely on temporary duty personnel for the other embassies. In my mind, this negated the core value of the SOC Forward, although I do not believe that the effectiveness of these regional constructs—which were more appropriately described as mini joint special operations task forces (JSOTFs)—has been closely studied. In SOCCENT, we employed this mini-JSOTF concept in our economy-of-force countries, when we dual-hatted the SEAL

"direct access to the country team and to the ambassador" and served as our "eyes and ears in country,"[11] but it also allowed the ambassador to execute the indigenous option under the authority of the Department of State as the SOC Forward was already part of the country team and could coordinate and execute this campaign on behalf of SOCCENT and CENTCOM.[12]

Additionally, each of the three campaigns was decidedly (and by necessity) different from one another, and the SOC Forward provided us the local context needed for customizing these campaigns. A whole-of-government flavor in these campaigns was unavoidable, because the ambassador's vision of the fight necessarily drove our own vision, and the SOC Forward allowed us to understand that vision, advocate for it within military channels, and ensure that we provided the necessary resources to achieve that vision. In each case, I engaged directly with the ambassador, the chief of station, and appropriate leadership of the host nation to validate these campaign plans. Ensuring that they were aligned with the objectives of both the ambassador and the geographic combatant commander was my number-one task and not something that could be delegated.

The potency of the SOC Forward construct in CENTCOM is perhaps best illustrated by SOCCENT's partnership with the Pakistani Frontier Corps.[13] In 2007, the Department of Defense decided that it wanted to improve the training offered to the Frontier Corps,[14]

---

command in Bahrain as JSOTF–geographic combatant command to work toward our operational objectives in the Gulf, primarily to prepare our partners for operations against potential Iranian threats in the event of war.

[11] Lehman, 2016, p. 6; Jack J. Jensen, "Special Operations Command (Forward)—Lebanon: SOF Campaigning 'Left of the Line,'" *Special Warfare*, April–June 2012, p. 29.

[12] Rob Newsom, "Adapting for the 'Other' War," *Small Wars Journal*, October 18, 2013.

[13] The Frontier Corps is also referred to as the Frontier Scouts.

[14] Ron Synovitz, "Pentagon Wants More Funding for Pakistan Frontier Corps," Radio Free Europe, November 20, 2007; Fawzia Sheikh, "DOD: 30 U.S., U.K. Personnel to Mentor Pakistan's Frontier Corps," *Inside the Pentagon*, January 10, 2008.

a force that U.S. Special Forces had worked with since 2003.[15] A key component of this new training was the construction of a very large training complex near the Afghanistan-Pakistan border. Ambassador Anne Patterson, who had fond memories of working with U.S. Special Forces during her time as ambassador to Colombia, gave SOCCENT oversight and responsibility not only for Frontier Corps training but also for the construction of its multimillion-dollar training compound. This role would give SOCCENT outstanding access to its units operating in the Federally Administered Tribal Areas.

This project was initially managed by SOCCENT via a special operations command and control element, a special operations element typically designed for coordination and synchronization with U.S. conventional forces.[16] Lieutenant General Mulholland had formed this element from a handpicked collection of experienced officers, noncommissioned officers, and support staff and asked Colonel Jeff Waddell to be its first commander. Colonel Waddell, a former commander of the South America–focused 7th Special Forces and thus familiar with our operations in Colombia, proved an inspired choice. During my tenure, he would masterfully manage the transition of his element into a SOC Forward,[17] which was purposely designed to more effectively integrate with the embassy while still maintaining a capability to reach back to SOCCENT for administrative, logistics, and intelligence support.

SOCCENT's management of this project, via these elements, gave us the capability to exploit the access that the construction of the facility allowed.[18] SOCCENT was able to rapidly assemble reservists with the right expertise for the mission (e.g., carpenters, plumbers, electricians) to support a small attached detachment of active duty

---

[15] William Rosenau, *"Irksome and Unpopular Duties": Pakistan's Frontier Corps, Local Security Forces, and Counterinsurgency,* Alexandria, Va.: CNA, May 2012.

[16] Joint Publications 3-05, 2003, p. xii.

[17] At the time, the *Special Operations Command and Control Element* was a doctrinal term, while the *SOC Forward* was not.

[18] Ambassador Patterson had allowed the establishment of a Special Operations Command and Control Element after arriving in Pakistan, which was transitioned into a SOC Forward after I took command of SOCCENT.

tradesmen, a very nonstandard but effective arrangement. The SOC Forward managed them as they rotated to a worksite to provide quality control in one of the most dangerous spots on earth. Yes, the Army Corps of Engineers could have handled the technical aspects of the construction of this facility. But access to and leverage with the Frontier Corps offered overt and low visibility options for the United States, options that SOCCENT could employ in support of U.S. strategic interests in the region. We were the only agency or command that had such expertise and could sustain a small enough footprint to avoid (for the most part) stirring local passions yet protect itself sufficiently and still accomplish the mission.

In the following years, SOCCENT built from this opportunity to deepen its relationship with the Frontier Corps across the Federally Administered Tribal Areas. In one case, we deployed a Special Forces medic and warrant officer into a Frontier Corps base, and that medic saved several lives when the base came under attack. In another, three U.S. Special Forces soldiers were killed in an attack targeting the Frontier Corps, at a ceremony celebrating the reopening of a girls' school previously destroyed by the Taliban.[19] These examples, and many more that would have been almost unimaginable in earlier years, built resilience for the first time in an uneasy and delicate relationship.

Mutual trust, however, with the Pakistanis would remain elusive.[20] For one, as it turned out, the temporary training site for the Frontier Corps was just blocks from the Pakistan Military Academy in Abbottabad. Thus, for nearly a year, while the permanent training facility was being built, Special Forces soldiers trained members of the Frontier Corps within blocks of Osama bin Laden's safehouse. Although this undoubtedly left Pakistan intelligence operatives pleased with their subterfuge, we were in retrospect not all that surprised: Special Forces operators on the Afghan side of the fight typically knew

---

[19] Nick Schifrin and Habibullah Khan, "3 U.S. Special Forces Die in Pakistan Bombing," ABC News, February 3, 2010.

[20] In the wake of the Osama bin Laden raid, relations soured to the point that U.S. Special Forces advisers with the Frontier Corps would soon be withdrawn, although there was little that we or the U.S. government could possibly have hoped to do to prevent that (Karen DeYoung, "U.S. Withholding Military Aid to Pakistan," *Washington Post*, July 10, 2011).

the Pakistani intelligence agents running or supporting Taliban groups in their area of operation, but there was little we could do to disrupt them. In the end, following the raid against Osama bin Laden, it can be argued that the good-faith efforts of SOCCENT with the Frontier Corps built resilience into the relationship between our militaries that tempered Pakistan's official anger.

Designing and Executing Irregular Warfare Campaigns for the "Big" Wars: The Value of Operational-Level Special Operations Headquarters

In 2008, when I assumed command at SOCCENT, there was a growing awareness among senior leaders that U.S. special operations in Afghanistan and Iraq were not being effectively employed in these two irregular warfare campaigns. Our special mission units in these theaters were being used to great effect in national-priority counterterrorism operations, but the predominantly conventional leadership in these two contests appeared to struggle in effectively employing special operations' indigenous-centric irregular warfare capabilities. During my tenure at SOCCENT, we would establish and support general officer–level special operations headquarters in both theaters, purpose-built to support the theater commanders in designing and executing irregular warfare campaigns that more effectively leveraged the unique capabilities of these special operators.

By the fall of 2008, a lack of progress in the Afghan campaign was beginning to create significant anxiety within the U.S. presidential administration,[21] and senior military and political leaders indicated that there was a "new urgency to put the mission in Afghanistan on the right path."[22] That summer, the Taliban had "regrouped . . . [and] coalesced into a resilient insurgency," and attacks against civilians,

---

[21] Paul D. Miller, "Obama's Failed Legacy in Afghanistan," *American Interest*, February 15, 2016.

[22] Eric Schmitt and Thom Shanker, "Bush Administration Reviews Its Afghanistan Policy, Exposing Points of Contention," *New York Times*, September 22, 2008.

Afghan security forces, and international forces operating in Afghanistan were on the rise.[23] Despite the size of the international military presence at the time, as both the United States and the international community had more troops in Afghanistan in 2008 than at any time earlier in the war,[24] and the effectiveness of those forces in tactical engagements against the Taliban, the security situation continued to worsen.[25]

At the time, the U.S. conventional military leadership in Afghanistan—particularly the rotating corps and division commanders—believed that this lack of progress was in part because of the misuse of U.S. Special Forces in the conflict. The central role of Special Forces in the toppling of the Taliban in 2001 was unquestioned. However, by 2008, there was a perception that U.S. Special Forces, who were working with various indigenous organizations inside and outside the government, were not effectively coordinating their operations and in some cases becoming a detriment to the broader U.S. strategy. Many of these conventional commanders also saw a need for a tighter relationship with various elements of the Afghan military and believed that this mission should be given to the language-qualified and combat-proven Special Forces units.

This perspective reflected a common misunderstanding about U.S. Special Forces. It is certainly true that each Special Forces team is designed to train and employ a battalion's worth of indigenous troops, and our teams had used this capability to great effect in Afghanistan. These partnerships proved critical against the Taliban, both during the initial invasion and in later years through informal partnerships with the Afghan military, police, militias, tribes, and warlords. However, Special Forces teams are designed primarily to build irregular forces and do not possess the wide variety of specialties necessary to create

---

[23] U.S. Department of Defense, *Report on Progress Toward Security and Stability in Afghanistan*, Washington, D.C., June 2008, p. 6.

[24] Ian S. Livingston and Michal O'Hanlon, *Afghanistan Index*, Washington, D.C.: Brookings Institution, May 25, 2017.

[25] Linda Robinson, *One Hundred Victories: Special Ops and the Future of American Warfare*, New York: PublicAffairs, 2013, pp. 12–14.

a modern army. Building a modern Afghan National Army beholden to the central authority in Kabul was just simply beyond the scope of what we could possibly hope to do.

We at SOCCENT, who at the time directed how U.S. Special Forces were being employed in Afghanistan, were willing and able to support U.S. efforts to build the Afghan National Army. Indeed, beginning in 2006, our teams began what would be a long-term and persistent effort to develop, train, and advise the Afghan National Army Commandos and Special Forces. But effectively building an Afghan National Army required that the U.S. Army develop and deploy trainers and advisers from across the conventional force. And we were concerned that any effort to divert our teams from their ongoing informal partnerships, which had proven one of the few effective tools against the Taliban, would result in the United States conceding even more terrain to the Taliban.

In hopes of addressing these concerns, then–Major General Mulholland had proposed a meeting between General David McKiernan (the senior commander in Afghanistan) and Admiral Eric Olson (the SOCOM commander). This meeting, which would address the concerns about both partnering and command and control, was held in October 2008 and thus three months after the SOCCENT change of command. As a result, I was to have the great fortune of representing SOCCENT, and hence the irregular warfare-focused component of the special operations community, in one of the most consequential leadership decisions in special operations' history in the war in Afghanistan.

Admiral Olson hosted this meeting, which we called a special operations *shura* (the Afghan word for a *council of respected leaders*[26]), at Bagram Airfield in Afghanistan.[27] Our hope was that we could forge

---

[26] The term *shura*, though used ubiquitously by U.S. military elements in Afghanistan, was perhaps particularly appropriate for this discussion, as *shuras* "[i]deally . . . represent the political groups within a given community," which was exactly what we were hoping to achieve (Shahmahmood Miakhel and Noah Coburn, *Many Shuras Do Not a Government Make: International Community Engagement with Local Councils in Afghanistan*, Washington, D.C.: U.S. Institute of Peace, September 7, 2010, p. 2).

[27] The meeting was held at the headquarters of the Combined Joint Special Operations Task Force–Afghanistan (CJSOTF-A).

a new approach for the use of U.S. Special Forces in country. The conversation was frank, and General McKiernan asked us hard questions about how the Special Forces assigned to the Combined Joint Special Operations Task Force–Afghanistan (CJSOTF-A), the tactical-level special operations headquarters in Afghanistan that SOCCENT controlled,[28] could be better employed. At one point, I offered to move a Special Forces team out of an area to allow for a formal partnership with a particular brigade, but General McKiernan quickly replied that he needed someone there. The team stayed in place. Our teams were deployed where others were not, and our relatively small force was preventing the Taliban from regaining critical territory even if our informal partnerships were not producing clear-cut victories or could be said to be formally building partner-nation capacity.

At the conclusion of the shura, General McKiernan agreed that a new approach was required.[29] Central to the new approach would be the establishment of a new general officer–led operational-level special operations headquarters in Kabul. This new headquarters would address Admiral Olson's concerns that special operations were simply not getting "into the room where the big decisions were made."[30] At the time, it was SOCCENT's responsibility to support General McKiernan in campaign development and execution.[31] But the reality was that both SOCCENT's liaison officer in Kabul and the colonel-level CJSOTF-A headquarters based at Bagram Airfield were ill postured to do so.[32] For CJSOTF-A, a lack of rank and the limited number of staff officers at this tactical-level command, combined with geographic separation

---

[28] CJSOTF-A was under SOCCENT's operational control, which meant that we determined where and how its force would be employed.

[29] Despite McKiernan's support for this approach, we would still face some subsequent animosity from some of the conventional division commanders.

[30] Robinson, 2013, p. 14.

[31] In 2008, SOCCENT had operational control of most U.S. Special Forces in Afghanistan, although that would change shortly after this shura.

[32] At this point, the CJSOTF-A headquarters consisted of a Special Forces group commander and staff.

from Kabul, where all the decisions were made,[33] meant that this head-quarters was simply not structured to provide input to the multitude of boards, bureaus, centers, and cells that ran America's half of this irregular war.

A one-star special operations headquarters subordinate to SOCCENT was my proposal, and to my pleasant surprise both four-stars agreed. The Combined Forces Special Operation Component Command–Afghanistan was stood up in January 2009, with Briga-dier General Edward Reeder, arguably the best Afghan hand in special operations, as its first commander. This new headquarters was to be the first-ever operational-level special operations headquarters centered on building and employing indigenous assets.[34] Based in Kabul, its pri-mary role was to interface with U.S., coalition, and Afghan leadership and to address Kabul-centric political issues related to special opera-tions efforts to develop the Afghan security forces.[35] It would lead the

---

[33] Bagram Airfield is approximately 30 miles from Kabul.

[34] Initially, the theater commander had tactical control of this headquarters, while SOCCENT maintained operational control. This ensured that the headquarters would sup-port the theater commander but leverage SOCCENT's knowledge of the campaign and synchronize efforts with operations in Afghanistan's neighbors. However, the theater com-mander assumed operational control in April 2010, forfeiting the expertise that SOCCENT offered and shifting control to a staff that had relatively limited experience with irregu-lar warfare (for a discussion, see Donald C. Bolduc, "The Future of Afghanistan," *Special Warfare*, October–December 2011). Although this move was justified under the auspices of improving unity of command, I know of no materially good reason for the change, nor could the theater commander give me one in our meeting on the subject. And I was not the only one, at least at that time, to be concerned that this move would reduce the effectiveness of our special operations assets in theater (Andrew Feickert, *U.S. Special Operations Forces [SOF]: Background and Issues for Congress*, Washington, D.C.: Congressional Research Ser-vice, July 16, 2010, pp. 7–8).

[35] Brigadier General Reeder set up an Afghan-style tearoom, where he held shuras with a wide range of Afghan officials and power brokers. Following a decision to bring both "national" (i.e., those focused on executing direct-action or strike missions) and "theater" (i.e., those focused on indigenous missions) special operations forces under one special opera-tions two-star headquarters, the commander converted this space into a situational aware-ness room, basically a command and control center. This was symbolic of the differences between the two very different halves of American special operations, with one very aligned to the American way of war, the other at the forefront of the underappreciated American way of irregular war.

fight for resources and mission for the men and women of CJSOTF-A and their Afghan allies. The intent was that this headquarters would also, eventually, take over tactical control of both the U.S. special mission units and other NATO special operations forces operating in Afghanistan.[36]

Alongside the establishment of this new headquarters, General McKiernan agreed that CJSOTF-A and SOCCENT should pilot a "local defense" initiative. By this time, it was clear that a new approach was needed to contest Taliban control of rural areas. Special operations had a long history of working with local communities to defend themselves against the Taliban and other insurgencies, such as in El Salvador and Vietnam, and the hope was that a similar bottom-up approach might be employed in Afghanistan.[37]

The first attempt at such a local defense program, the Afghan Public Protection Program, was in place just a few months later. Initially, there was substantial resistance to these efforts, with critics concerned that it would empower local warlords or tribal leaders and further weaken the central government. However, much of this criticism was from Western voices, and many Afghans recognized the value of such an approach if appropriately executed. Eventually, in August 2010, Afghan President Hamid Karzai formally approved a subsequent initiative, known as the Afghan Local Police, that would combine local legitimacy and connectivity to the national government.[38] Over the

---

[36] Its successor, Special Operations Joint Task Force–Afghanistan, ultimately achieved only partial success in bringing together command and control over all special operations forces—the belief that the national force needed the ability to immediately move assets when they deemed necessary proved to have powerful advocates.

[37] I called this *reverse bridging*. The idea was instead of a top-down imposition of security and forced fealty to Kabul, build trust at the local (often tribal) level first, then bridge loyalty to the district, then maybe the province, and ultimately Kabul. But if Kabul was a bridge too far, you could satisfy your (U.S.) security requirements with simply local allegiance or tacit control.

[38] The defining characteristic of the Afghan Local Police was an emphasis on local legitimacy and connectivity to the national government. The Afghan Local Police was similar to its predecessor initiatives in that it was special operations–led with special operations teams embedding in rural communities to simultaneously improve governance, development, and security.

next four years, U.S. special operations would lead the development of a force of nearly 30,000 "local guardians" before the program was formally transitioned to the Afghan Ministry of Interior Affairs.

Within a week of the agreement to establish this new special operations headquarters in Afghanistan, the theater commander in Iraq asked both me and Admiral Olson where "his" special operations general was. From my perspective, this suggested that General Raymond Odierno, who had just (in September 2008) assumed command of Multi-National Force–Iraq, recognized that his brigade, division, and even corps staffs lacked the expertise to effectively employ the indigenous-centric irregular warfare capabilities of the special operations units under his control. Seeing us build an architecture that would allow General McKiernan to optimize the use of special operations in achieving his objectives in Afghanistan, General Odierno asked that SOCCENT do the same in Iraq.

The establishment of an equivalent operational-level headquarters in Iraq would take a bit longer than in Afghanistan, because much of the Multi-National Force–Iraq leadership was less receptive of enhanced special operations leadership than their commander.[39] Indeed, it rankled many that the existing colonel-level special operations headquarters, the Combined Joint Special Operations Task Force–Iraq, had been operating largely independently.[40] So, we first established a general officer–level special operations directorate within the theater command staff, the Special Operations Directorate–Iraq, which allowed us to keep peace with General Odierno's staff and that of XVIII Airborne Corps that was managing the fight for the theater. We then transitioned this staff directorate into an operational-level

---

[39] Special operations and conventional forces had historically struggled to effectively integrate their efforts with calls for the special operations forces to be subordinated to local conventional commanders (Catherine Dale, *Operation Iraqi Freedom: Strategies, Approaches, Results, and Issues for Congress*, Washington, D.C.: Congressional Research Service, March 28, 2008, p. 59). It is a testament to the theater commanders in Iraq that they did not succumb to those calls for such control.

[40] Combined Joint Special Operations Task Force–Iraq was operational control to SOCCENT and tactical control to the theater commander, which gave the corps limited authority over the U.S. special operations forces operating in theater.

special operations headquarters, the Joint Forces Special Operations Component Command–Iraq, a year later, in late 2009.

Both of these constructs, both the staff directorate and the operational-level headquarters, allowed the special operations community to provide campaign planning input to the theater command. They also oversaw U.S. support to the Iraqi Counter Terrorism Service, which included mentoring and advising, direct training, and train-the-trainer efforts, from 2008 to 2011.[41]

Unlike in Afghanistan, SOCCENT maintained operational control of these operational-level structures in Iraq, which proved critical when the U.S. mission in Iraq ended abruptly in December 2011. SOCCENT, facilitated by its in-country headquarters, was able to ensure that selected indigenous assets were either retained or transitioned to other appropriate agencies. Foremost among the indigenous assets protected was the Counter Terrorism Service, which SOCCENT transitioned to the CIA and Office of Security Cooperation–Iraq and thus ensured effectively no break in U.S. mentorship.[42] This turned out to be a wise investment, because, beginning in 2014, the Counter Terrorism Service would prove critical in international efforts against the Islamic State.[43]

As the wars in both theaters evolved, these operational-level headquarters proved critical in both planning for operations and in helping prepare incoming units for new roles, missions, and locations. In addition, they provided a new capability to coordinate the concept development, organizations, training, and support needed to stand up institutions that could produce and support the indigenous special operations units that the United States had created. Although the Joint Forces Special Operations Component Command–Iraq built the operational

---

[41] Richard R. Brennan Jr., Charles P. Ries, Larry Hanauer, Ben Connable, Terrence K. Kelly, Michael J. McNerney, Stephanie Young, Jason H. Campbell, and K. Scott McMahon, *Ending the U.S. War in Iraq: The Final Transition, Operational Maneuver, and Disestablishment of United States Forces–Iraq*, Santa Monica, Calif.: RAND Corporation, RR-232-USFI, 2013, p. 187.

[42] David Whitty, *The Iraqi Counter Terrorism Service*, Washington, D.C.: Brookings Institution, 2016, pp. 24–25.

[43] Whitty, 2016, pp. 35–37.

and sustainment capabilities of the Counter Terrorism Service in Iraq, the Combined Forces Special Operations Component Command–Afghanistan (and its two-star successor) helped create the Afghan National Army Special Operations Command that would build, train, and sustain the constellation of Afghan commando and special forces units.[44]

The potency of these two operational-level special operations headquarters is, perhaps, best demonstrated by the central role that their indigenous partners continue to play in the wars in both Afghanistan and Iraq. In Iraq, after the Iraqi Army was routed in 2014, it was the Counter Terrorism Service and Iraqi special operations forces that prevented the Islamic State from entering Baghdad and that later conducted the bulk of the combined operations to defeat the Islamic State. Similarly, it is the Afghan National Army Commandos and Special Forces that are the lead strike and response capability for the Afghan National Army.[45] But this outcome is unsurprising, as it is irregular warfare in which the indigenous warfighter is necessarily the centerpiece. After all, it's their country.

Organizing for Population-Centric Conflict: The Human Domain

It was during a speech at the Jordanian Special Operations Exposition in 2010, my final year as the SOCCENT commander, that I first publicly articulated a concern that had been nagging at me for at least a decade. By this time, it was clear that our approach in contests across the globe, and in the CENTCOM area in particular, was not working. The big wars in Afghanistan and Iraq had become quagmires, despite our tremendous initial success in each, and we were steadily losing

---

[44] The successor to the Combined Forces Special Operations Component Command–Afghanistan was the two-star Special Operations Joint Task Force–Afghanistan, which was established in 2012.

[45] Their conventional counterparts trained by other U.S. elements, in comparison, suffer chronic desertion and corruption.

ground in many of the conflicts below the threshold of traditional conflict (e.g., against Hezbollah in Lebanon).

The fundamental difficulty as I saw it then, and I believe the observation remains true today, is that the domain model that the United States uses for organizing its approach to national security is not well designed for population-centric warfare. There are now five (then only four) recognized domains—air, cyber, land, sea, and space—with each of the services contributing to two or more different domains,[46] and the domain model provides "the US military with a division of labor and knowledge for creating, developing, and employing warfighting capabilities."[47]

The problem is that these domains were designed for a different adversary from the ones we face in population-centric contests. Many have argued that these conflicts are part of the land domain. But can the same domain model really provide us a credible capability for mechanized warfare against a near-peer adversary, counterterrorism operations that rely on speed and surprise,[48] hybrid warfare operations that typically require a decisive and overwhelming response,[49] and the sometimes decade-long, indigenous-centric campaigns often required for population-centric conflicts?

The reality is that many of today's contests are happening in a human domain, a warfighting domain for which we do not have the right tools for success. We are thus failing, at least in part, because our approach does not properly account for the fact that these contests are about the indigenous people. These are contests of will, and we must make our case or create support for our cause, primarily through an

---

[46] Michael Spirtas, "Toward One Understanding of Multiple Domains," *RAND Blog*, May 2, 2018.

[47] Robert L. Cornelius, *An Evaluation of the Human Domain Concept: Organizing the Knowledge, Influence, and Activity in Population-Centric Warfare*, Fort Leavenworth, Kan.: School of Advanced Military Studies, U.S. Army Command and General Staff College, 2015.

[48] William H. McRaven, *Spec Ops: Case Studies in Special Warfare Operations: Theory and Practice*, New York: Presidio Press, 1995, p. 1.

[49] King Mallory, *New Challenges in Cross-Domain Deterrence*, Santa Monica, Calif.: RAND Corporation, PE-259-OSD, 2018.

indigenous host so as to avoid the trap of an occupier, if we hope to achieve enduring strategic success.[50] As we have learned the hard way in our population-centric wars since the Vietnam War, contorting the conventional instruments of war built for the land domain are insufficient, and often detrimental, to achieving strategic success in any reasonable time frame and within acceptable costs.[51]

A central challenge, therefore, is that neither the Department of Defense nor any other organization within the U.S. national security enterprise has the concepts and resulting operational-level structures for developing and conducting the campaigns necessary in these population-centric conflicts. The deficiency is not at the tactical level, as recent history further demonstrated that U.S. traditional and irregular warfare capabilities are very capable and can adapt rapidly. What is lacking is the conceptual thinking at the operational level to orchestrate or support indigenous combat capabilities, psychological operations, governance assistance, and economic development without the employment of large U.S. formations. Where physics controls land domain warfare, psychology dominates in the human domain.

Within the Department of Defense, and the Army more specifically, there has not yet been a deliberate effort to build a land domain construct for population-centric warfare comparable to that for Air-Land Battle or multidomain operations. No current headquarters or organization has as its core function the folding of U.S. technical support, fires, and intelligence around an indigenous mass. As a result, or perhaps as a consequence, our doctrine and broader capabilities are too immature and too U.S.-centric.[52] The consequence of this has been especially evident at CENTCOM, where planners have seemed unable to anticipate the long-term negative effects of using U.S. conventional forces in place of indigenous forces, and host-nation capacity-building

---

[50] Odierno, Amos, and McRaven, 2013.

[51] See, e.g., Jeffrey Record, *The American Way of War Cultural Barriers to Successful Counterinsurgency*, Washington, D.C.: CATO Institute, 2006.

[52] With *broader capabilities*, I am explicitly referencing organization, training, materiel, leadership and education, personnel, facilities, and policy, which, along with doctrine, constitute DOTMLPF-P.

efforts have been complicated by mirror imaging and lofty nation-building objectives.[53]

Success in population-centric fights requires that the United States take its partners to the bloody edge of success but resist going beyond that. It has to be their fight more than ours, and we have to resist doing too much for them. In this form of warfare, American combat troops have to be used sparingly, if at all; U.S. expectations have to be tempered; and a realistic long view must be taken. Social change, even more than economic development, has a pace of its own, and an outside power risks campaign failure if it tries to artificially impose an unrealistic timeline on changes to societies. Ensuring security is still the initial priority, although that security is achieved through a strong host-nation central government, a loose collection of strongmen, or another indigenous solution rather than via an armed occupation.

What success in the human domain requires is the development and maturation of an American way of irregular war, which I would later realize is also critical in supporting today's global political warfare competition. Yet I would learn during my time at SOCCENT that there was no real proponent for this form of warfare within the Department of Defense.

SOCOM, which was on paper responsible for this form of warfare, simply did not see the development of our irregular warfare capabilities as a priority. Some saw it as a zero-sum game, as the development of this capability would come at the cost of SOCOM's manhunting capability; others were resistant to support any initiatives that might disrupt the status quo and did not want to get more engaged in the stalemates in Afghanistan and Iraq; and others thought that these conflicts ought to be the domain of either the CIA (for the irregular stuff in small conflicts) or the Army (for the big ones).[54] This lack of interest in irregular

---

[53] Substantial Foreign Military Sales and U.S. development assistance were used to make these disruptive changes more palatable but did not make achieving these objectives any more feasible.

[54] Although the 9/11 Commission Report had recommended that all U.S. paramilitary operations be shifted to SOCOM, this recommendation was rejected by the CIA (Richard A. Best Jr. and Andrew Feickert, *Special Operations Forces [SOF] and CIA Paramilitary Operations: Issues for Congress*, Washington, D.C.: Congressional Research Service, 2006).

warfare was illustrated vividly, at least to me, by SOCOM's apparent indifference to not being included in the much-ballyhooed writing and rewriting of Field Manual 3-24, the doctrine guiding our approach in Afghanistan and Iraq.[55] Overcoming this indifference would dominate much of my time at my next assignment, at U.S. Army Special Operations Command.

---

[55] Field Manual 3-24 and Marine Corps Warfighting Publication 3-33.5, *Insurgencies and Countering Insurgencies*, Washington, D.C.: U.S. Department of the Army, May 2014.

CHAPTER ELEVEN

Shepherding America's Irregular Warfare Capability

I took command of U.S. Army Special Operations Command (USASOC) in July 2012. America was actively fighting insurgents and terrorists (both nonstate and state-sponsored) across the globe, and U.S. special operations forces were taking the fight to the enemy in Africa, Asia, Europe, Latin America, and the Middle East. This made sense, as these were essentially irregular campaigns, although they were not necessarily understood as such.[1] They were campaigns for which special operations was purposely designed, and they meant that USASOC would stay on a war footing throughout my tenure. USASOC was responsible for just over 50 percent of the total manpower for U.S. special operations and 80 percent of the "operator" strength for U.S. Special Operations Command (SOCOM).

More importantly, from the perspective of irregular warfare, was that USASOC was responsible for nearly all of the nation's capacity dedicated to working with indigenous forces, both state and nonstate. USASOC was also the home of the John F. Kennedy Special Warfare Center and School, the premiere center in the United States for thinking about irregular warfare. We were, essentially, the military arm of American irregular warfare.

I went into command with the belief that USASOC needed to build a world-class irregular warfare capability that, like America's

---

[1]   This was perhaps particularly the case for the largely conventionally minded geographic combatant commands.

peerless direct action capability, was purpose-designed to operate at the operational and strategic levels. This capability would be built around U.S. Special Forces, psychological operations, and civil affairs units, the USASOC forces dedicated to this form of warfare. Although this view was admittedly colored by my own career, I also wanted to change the narrative about U.S. special operations, as I found that there was often ignorance, both within and outside special operations, about the importance of "my" form of warfare. In short, my central responsibility, as I saw it, was to ensure that USASOC became the shepherd of America's irregular warfare capability. It was pretty much an open field.

It was the support of General Ray Odierno and Admiral William McRaven that would give me the ability to move aggressively toward this vision. As the USASOC commanding general, I had two bosses: the chief of staff of the Army and the commander of SOCOM. General Odierno, who would hold the chief of staff position throughout my tenure, was my "blue-collar dad" who controlled the institutions and processes that governed how we got soldiers, organized and educated ourselves, and maintained order and discipline. Admiral McRaven, who commanded SOCOM for most my tour at USASOC, was "my rich stepdad," whom I would need to underwrite the changes that I saw we needed. Both of these men, who were among the most open and receptive I ever worked for, recognized the value of what we were doing at USASOC. Their support would prove critical.

We set an ambitious agenda, focused on increasing readiness, establishing a deployable division-equivalent headquarters designed for executing irregular warfare campaigns, formalizing some of the concepts that we would need to mature our irregular warfare capability, and building institutional structures critical to the success of this capability. In the first two weeks after assuming command, we addressed a handful of major decisions that my predecessor had put on hold until the change of command,[2] and then turned ourselves to pre-

---

[2] This included committing to replacing our fleet of aging CASA-121s with some new aircraft being scrapped by the Air Force, requiring all Special Forces soldiers to be freefall qualified, and reorganizing USASOC's Army Compartmented Element.

paring USASOC to be the proponent for American irregular warfare, while balancing our duties as the caretaker of its premier commando capability. My council of gray beards was quick to tell me how difficult these changes would be. I admit that I took this pessimism as a challenge, as this group generally agreed with the ideas, just not the chances of success.

During my time at USASOC, we were able to take some major doctrinal steps forward. Foremost among these was the publication of the first Army doctrine for the employment of special operations above the tactical level. Although much of the groundwork was done well in advance of my arrival at USASOC, I had the opportunity to direct its content and approve its final development during my ten-month "apprenticeship" between my Senate confirmation and when I assumed command in summer 2012. This doctrine was published by the Army shortly after I took command and proved critical for legitimizing the approach that we would take for reforming Army special operations during my tenure at USASOC.

Further, in what amounted to bureaucratic jiujitsu, we used the unprecedented resource disruption created by the Budget Control Act of 2011 to drive rapid significant organizational change.[3] In under three years, USASOC's staff and its subordinate headquarters created both a dedicated unconventional warfighting capability and a deployable, operational-level irregular warfare capability, while at the same time paying its quarter-billion-dollar "sequestration tax." In retrospect, the hallmark of my time at USASOC was using the looming sequestration to create the opening for the unprecedented changes that had been proposed.

We knew that redesigning USASOC was a critical but insufficient step in creating the irregular warfare capability that America needed and that a world-class irregular warfare capability would require partnerships throughout the Department of Defense and across the U.S. government. Part of our outreach was via the Campaign of Learning, a USASOC-driven future-exercise program that we created to engage with the myriad military, civilian, academic, and industry elements

---

[3] Pub. L. 112-25, Budget Control Act of 2011, August 2, 2011.

that had a role in irregular warfare or other special operations activities. But it was the support of General Odierno and Admiral McRaven, who recognized the broader value of what we were doing, that would drive several Department of Defense–wide efforts that were critical in enhancing America's irregular warfare capability.

What we were able to accomplish was a testament to the exceptional men and women assigned to that command. And our efforts would bear fruit in the years that followed, driving meaningful discussion and debate across the national security community and even precipitating some modest changes in the American approach to irregular warfare.

But much would remain undone. I learned at the end of three years that the type of reform necessary to mature and truly professionalize the American way of irregular war would require that the United States address some fundamental problems in our national security models while developing new capabilities to respond to the rapidly changing information environment and increasingly interconnected world. These changes were beyond the scope of even the most senior commanders within our military.

Developing Doctrine and a Ten-Year Vision for Army Special Operations: A Foundation for Reforming America's Approach to Irregular Warfare

The Senate confirmed my appointment to command USASOC in the summer of 2011, but I would not take command for another ten months, in July 2012. During this ten-month "apprenticeship" or "sabbatical" in advance of assuming command, I had the unique opportunity to study the command in depth, get to know its leaders, meet my peers across the Army, and firm up my vision for Army special operations.

I came into USASOC believing that it was the best qualified among all U.S. government agencies and other military commands to take the lead in developing the concepts, doctrine, and personnel necessary to meet the requirements for the military segment of America's irregular warfare arsenal. USASOC's soldiers had been the brains and brawn of America's tactical-level irregular warfare capabilities since well before its formation in 1989, and the planners, thinkers, and writers at USASOC's John F. Kennedy Special Warfare Center and School gave the command a good start on the capability needed.

This belief was vindicated during the early days of my apprenticeship, when I had the unique opportunity to be part of the writing and editing of what would become Army Doctrine Publication 3-05, *Special Operations*. This unprecedented opportunity was a result of the trust and confidence that Major General Bennet Sacolick, the commander of the John F. Kennedy Special Warfare Center and School, had established with Army senior leadership. He had convinced key leaders that the existing Army doctrine was incomplete because of its limited and tactical treatment of special operations,[4] as well as that the John F. Kennedy Special Warfare Center and School should draft an Army doctrine publication framing how Army special operations fit into the joint fight. The hope was that this document would become one of the core publications that articulate how the Army and its components fight and win the nation's wars, most of which in the modern era have had significant involvement by Army special operations.

Army Doctrine Publication 3-05 would provide the first-ever holistic concept for the employment of Army special operations at the campaign level and in support of U.S. strategic objectives. The central component of this new doctrine was the reintroduction of the concept of special warfare.[5] We defined *special warfare* in Army Doctrine Publication 3-05 as the "combination of lethal and nonlethal actions" by a force with the "ability to build and fight alongside

---

[4] Field Manual 100-5, *Operations*, Washington, D.C.: Headquarters, Department of the Army, June 1993.

[5] The term *special warfare* was used in the 1960s by the Army as an umbrella term to describe the "military and paramilitary measures and activities related to unconventional warfare, counterinsurgency, and psychological warfare" (Elvis J. Stahr Jr., "Foreword," in John F. Kennedy, *Special Warfare*, Washington, D.C., Office of the Chief of Information, Department of the Army, 1962, p. 5). And this term was used by John F. Kennedy in his April 11, 1962, letter to the Special Forces community (John F. Kennedy, *Special Warfare*, Washington, D.C.: Office of the Chief of Information, Department of the Army, 1962, p. 3).

indigenous combat formations in a permissive, uncertain, or hostile environment."[6] This formalized a long-standing core competency of Army special operations—specifically, self-contained units with the specialties needed to build, train, and employ a wide range of indigenous, local solutions and conduct small unit raids behind enemy lines.

Special warfare units are those designed for sustained, small-footprint operations in austere environments. Their lethality and proficiency in small-unit warfighting, ability to rapidly assess situations, and acumen for relationship building makes them ideal for long-duration missions and particularly those in population-centric conflicts where host-nation solutions are desired. "Go slow, go long, go small, and go local" captured the special warfare approach. It required language training, understanding of culture, and an empathy for the plight of the oppressed and foreign defenders of freedom—the same freedom that Americans enjoyed.

This new doctrine also provided a venue for articulating the complementarity of Army special operations' two core competencies. On the one hand is special warfare, which provides the United States a capability to work with or through indigenous forces or surrogates. On the other hand is what Army Doctrine Publication 3-05 defines to be *surgical strike*, which encompasses all special operations executed "in hostile, denied, or politically sensitive environments" with the intent to "seize, destroy, capture, exploit, recover or damage designated targets, or influence threats."[7] *Surgical strike* thus refers to what Admiral William McRaven had, decades earlier, described as special operations' unique capability to "defeat a much larger or well-entrenched opponent" through the application of "speed and surprise."[8]

---

[6] Army Doctrine Publication 3-05, *Special Operations*, Washington, D.C.: Headquarters, Department of the Army, August 2012, p. 9.

[7] Army Doctrine Publication 3-05, 2012, p. 9.

[8] McRaven, 1995, p. 1. McRaven concludes that there are six principles—simplicity, security, repetition, surprise, speed, and purpose—that allow small strike teams to gain relative superiority and defeat these otherwise superior foes.

Army Doctrine Publication 3-05 made it clear, perhaps for the first time, that the "nation needs a world class capability in both."[9] The previous terminology used to describe these two components of America's special operations capability—*national* and *theater, black* and *white*, or *direct* and *indirect*—was at best vague,[10] and at times pejorative toward the practitioners of special warfare.[11] This, for me, was a first step in remedying the long-standing and unhealthy bifurcation of Army special operations.[12]

This new doctrine was rapidly approved by the Army, and Army Doctrine Publication 3-05 was published in August 2012, just one month after I took command. Sixty years after Army special operations was first established with the stand-up of 10th Special Forces Group in 1952, we finally had doctrine for Army special operations that went beyond tactical employment.[13]

Although this doctrine legitimized the approach that I would take for reforming Army special operations during my tenure at USASOC, the culmination of the ten months I spent awaiting command was a set of ideas that we would articulate with USASOC's first "annual

---

[9]   Charles T. Cleveland, James B. Linder, and Ronald Dempsey, "Special Operations Doctrine: Is It Needed?" *Prism*, Vol. 6, No. 3, 2016.

[10] Robinson, 2013, p. 11.

[11] Since my time as a battalion commander, this language had often been used to promote a narrative that those practicing special warfare were the "junior varsity" of the special operations community, while those who focused primarily on unilateral lethal missions were more skilled—when in reality they are expert in only a segment of what is broadly called *special operations*.

[12] This proved to be somewhat controversial, as it was criticized primarily by the raiding or surgical strike community for creating an unnecessary and unhelpful division within the special operations community—the opposite of the intended effect. Later versions of Army Doctrine Publication 3-05 would return to the less precise, and in my view more corrosive, language.

[13] The Chairman of the Joint Chiefs of Staff issued the first doctrine for joint special operations in 1992, with the first publication of Joint Publication 3-05, *Doctrine for Special Operations*, which was later revised on several occasions (Jerome Lynes, "A Critique of 'Special Operations Doctrine: Is It Needed,'" *Prism*, Vol. 6, No. 4, 2017; Joint Publication 3-05, 2013). The publication of Army Doctrine Publication 3-05 was the first time that Army-specific doctrine for special operations was drafted.

report" that we called ARSOF 2022.[14] This report provided the vision for change during my three-year tenure in command, describing the intellectual underpinnings of these changes and offering a positive and optimistic future that would fix chronic problems and increase the effectiveness of the command in its core contributions to defense of the nation.

The foundational concept of ARSOF 2022 was that USASOC, despite making up only 6 percent of the Army's active duty strength, was responsible for a unique part of the Army's operational spectrum. This concept was encapsulated in a single illustration,[15] reproduced in Figure 11.1, which summarized many of the controversial ideas that I had been wrestling with for more than a decade. The core argument of this illustration is that there is a segment of the Army's range of military operations for which Army special operations are the primary maneuver capability. USASOC, as the higher headquarters for these units, therefore needed to be the proponent for these forms of warfare within the Army. The figure also formalized the controversial concept of the human domain as the dominant domain, shared with the physics-focused land domain, in which these specific operations occurred.

ARSOF 2022 came at a critical time, as it also provided a format for communicating how USASOC would meet its obligation for the continued wars overseas in the face of the coming sequestration. At the time, there was great anxiety about the implications of the Budget Con-

---

[14] USASOC, "ARSOF 2022," special issue, *Special Warfare*, April–June 2013. *ARSOF* stands for *Army special operations forces.*

[15] I affectionately referred to this illustration as the "Cardona slide." Then–Major Gil Cardona, who was the first person assigned to help with in-processing and then stayed on as my aide de camp as the ten-month wait wore on, proved to be the perfect sounding board for the host of ideas that would find themselves in the ARSOF 2022. He had come from 18th Airborne Corps, where he had transitioned from a field artilleryman into a simulations officer, and provided a much-needed ability to look at issues without partisan bias and helped me create many of the models and diagrams to help communicate my vision. His replacements from Special Forces, the 160th Special Operations Aviation Regiment, and the Ranger Regiment would each play a key role in generating and vetting ideas, as my time as USASOC commander moved on, but Gil's candor, energy, and humor were absolutely critical in getting the ball rolling.

Figure 11.1
Core Competencies of Army Special Operations (the "Cardona Slide")

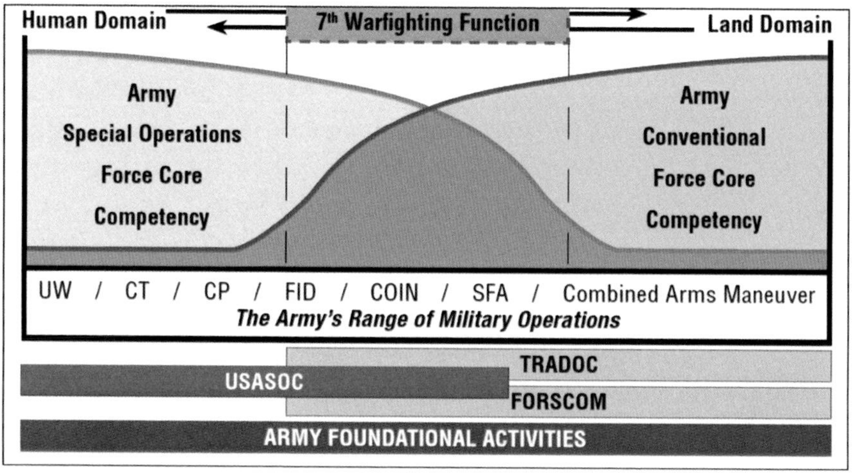

SOURCE: USASOC, "ARSOF 2022: Part II," special issue, *Special Warfare*, July–September 2014, p. 5.
NOTE: UW = unconventional warfare; CT = counterterrorism; CP = counterprolifera-tion; FID = foreign internal defense; COIN = counterinsurgency; SFA = security force assistance; TRADOC = U.S. Army Training and Doctrine Command; FORSCOM = U.S. Army Forces Command.

trol Act of 2011 for USASOC, as civilian layoffs, hiring slowdowns, civilian and military force reductions, and rollbacks in force structure were rumored and under way elsewhere. This report, therefore, delib-erately mimicked the goals of annual reports produced by major cor-porations, of which USASOC was certainly one, in communicating the status of the organization to its employees and "shareholders"— specifically, the American people through their representatives in Congress.

We published ARSOF 2022 in April 2013, some nine months after I took command. By the time that it was published, the redesign of USASOC that it suggested was already well under way. But it was this seminal document that provided the foundation for the types of capabilities that Army special operations needed and how these capabilities could be developed—all while paying the sequester bill.[16]

The engine behind the intellectual content in this seminal document, although based on my vision for USASOC, was the handpicked group of branch-qualified majors in my Commander's Initiatives Group.[17] Although the Budget Control Act of 2011 and the shrinking bottom line provided the necessary urgency to make change happen, this group and its partnership with the professional doctrine writers, trainers, and publishers at the John F. Kennedy Special Warfare Center and School gave us the tools needed to effect change—and prove the gray beards wrong.

We would publish two more annual reports during my tenure at USASOC. The second installment, ARSOF 2022: Part II, communicated how the institutional side of USASOC would have to change to support the new operational force, the centerpiece to the previous year's ARSOF 2022. It provided our rationale for reorganizing USASOC's approach to talent management. Ensuring that we delivered forces that were properly organized, trained, educated, and equipped for emerging operational requirements required that we make new investments in the human capital of our soldiers and develop an operational-level

---

[16] ARSOF 2022 focused on five core priorities: (1) Invest in human capital through education, training, recruitment, and supporting soldiers and their families; (2) strengthen relationships among special operations forces, conventional forces, the interagency, and international organizations; (3) support theater special operations commands by operationalizing existing regional expertise from within the U.S. government, academia, and industry; (4) develop special operations capabilities at the operational level; and (5) prepare deployable and scalable special operations command and control nodes to support all echelons of command.

[17] The Commander's Initiatives Group was complemented by a junior executive development program—which I called the Young Lions—that would bring one branch-qualified major from each of the nine tribes to USASOC headquarters for one five-day workweek. They were briefed on issues, went through leadership and team-building exercises, socialized, and then were given a relevant problem and told to come up with a solution that they would brief to the Commander's Initiatives Group. Their names were engraved on a plaque to permanently stay in the headquarters, and they were given access to the Commander's Initiatives Group portal that they would forever be able to use. They were subject to tasking after returning to home station and had to provide opinions on work being done in the group. They were the "classic" directed telescopes into our units.

irregular warfare capability. This volume also introduced our Campaign of Learning, which would include USASOC's first-ever future war game series (Silent Quest) and U.S.-based unconventional warfare field training exercise series (Jade Helm),[18] which was needed to close the gap with the parts of the interagency that were essential to irregular warfare.[19] The final report, "ARSOF Next: A Return to First Principles,"[20] reviewed our progress in building the irregular warfare capabilities that our nation required, concluding with the USASOC soldier's promise to the nation: "I protect the nation; without fear, without fail, without equal."[21]

Redesigning USASOC's Irregular Warfare Capabilities: Sequestration as a Forcing Function

We would shed nearly one-fourth of the total authorized Special Forces and Ranger force strength to pay our "sequestration bill" but used the uniqueness of the times to restructure USASOC to be a more effective headquarters for managing America's irregular warfare capability. Indeed, the hallmark of my time at USASOC was using the looming sequestration to create the opening for the unprecedented changes that we had proposed.

In December 2012, six months after taking command, we learned that USASOC would be asked to pay half of SOCOM's sequestration bill and find some $260 million in total savings by reducing force structure.[22] The cuts were significant, and we would shed planned

---

[18] Jade Helm was a Special Forces group-level exercise.

[19] The robust interagency network of SOCOM's three-star operational headquarters had proved to be a potent force multiplier in responding to hostage situations and other terrorist events, but no comparable capability existed for irregular warfare.

[20] USASOC, "ARSOF Next: A Return to First Principles," special issue, *Special Warfare*, April–June 2015.

[21] This motto was developed by my Commander's Initiatives Group, which I challenged to find a motto that captured the USASOC soldier while staying within 13 words.

[22] Drew Brooks, "Lt. Gen. Cleveland: Changing World Requires New Mission for Special Operations," *Fayetteville Observer*, March 18, 2015.

growth in the combat support capabilities of our Ranger and Special Forces formations, with each Ranger battalion losing a company and Special Forces losing nearly 1,000 soldiers (out of a total of around 8,000). However, other than the loss of support soldiers, the bite of these reductions was limited, as these cuts came from planned growth that both the Rangers and Special Forces had struggled to fill.[23]

Critically, these cuts provided a forcing function for the reforms that I believed necessary to bolster America's irregular warfare capability. The most radical of these reforms, the 4th Battalion redesign that transformed one battalion in each Special Forces group into a dedicated battalion-sized unconventional warfare capability, was also the reform that allowed me to pay the Special Forces component of my sequester bill, as the redesigned battalions would be half their original size. That we could do more with less and had a plan earned us support within SOCOM for our other proposed reforms. This created the needed opportunity to build a division-sized operational headquarters for irregular warfare, which we would call 1st Special Forces Command (Airborne), and develop the structures that would position USASOC to become the leader in the military science of irregular warfare.

The most radical tactical reform of the USASOC redesign was the development of a battalion-sized unconventional warfare capability in each Special Forces group. This redesign restructured the recently formed 4th Battalion of each group to be dedicated exclusively on unconventional warfare. These battalions would now be focused on conducting operations during the early stages of resistance, establishing an in-depth understanding of and, where possible, access in prior-

---

[23] The 2006 Quadrennial Defense Review had recommended the addition of a fourth battalion to each of the five active duty Special Forces groups and a new company for each of the three Ranger battalions, as well as a fourth aviation battalion for the 160th Special Operations Aviation Regiment (Linda Robinson, Austin Long, Kimberly Jackson, and Rebeca Orrie, *Improving the Understanding of Special Operations: A Case History Analysis*, Santa Monica, Calif.: RAND Corporation, RR-2026-A, 2018, Chapter 8). However, both the Rangers and Special Forces had had difficulty filling the additional allocations, so these cuts were actually relatively painless to the command. USASOC, 2014, p. 6.

ity areas even years in advance of potential military operations,[24] with the ability to operate in denied areas. This would ensure Special Forces readiness and capability to conduct unconventional warfare and support the CIA anywhere in the world.

These redesigned battalions are composed of three very different types of companies. The first is a collection of "singleton" Special Forces operators trained as case officers and with significant experience in their theaters of operations. These operators have been in high demand in recent years supporting interagency partners and will likely continue to be so for years to come. A second company is composed of 12 three-man Jedburgh teams, designed to be the first to deploy into potentially denied areas either to support another government agency or as a pilot team before the arrival of other elements from the Special Forces group. Mirroring the approach that we developed with 10th Special Forces' Jedburgh teams (see Chapter Eight), these teams are manned with native speakers (or as close as we could find to native speakers) who stay in a state of perpetual training, improving language skills, area familiarization, and regionally relevant sabotage and subversion techniques.[25] The third company provides command and control and support of these assets while deployed.[26]

---

[24] As a purely hypothetical example, a Jedburgh team from 7th Special Forces might have spent the past five years (since we formed these unconventional warfare battalions) focused on a single major city in Venezuela, getting to know the personalities, understand the local lingo, and become a Venezuelan. And then when an opportunity came up to add a driver to the local consulate, one of these people might become that driver. This would create opportunities: If the place falls apart and when the political decision has been made, we will be in a position to support. This could be part of an interagency effort or a military effort. This was, of course, a bit antithetical to how many think of Special Forces, but the nature of the war should drive how you prepare for it. In these circumstances, you need to specialize to get an advantage.

[25] Our Jedburghs differed from those used in World War II, when the concept was first developed, in that the teams used in World War II were multinational and included "at least one Frenchman since they were expected to operate inside France" (Tim Ball, "Bringing the Alliance Back to SOF: The Role of NATO Special Operations Headquarters in Countering Russian Hybrid Warfare," in *2016 Special Operations Essays*, MacDill Air Force Base, Fla.: Joint Special Operations University Press, 2016).

[26] For a discussion of this third company, see Andrew Basquez, "SF Returns to Its Roots with the 4th Battalion Redesign," *Special Warfare*, October–December 2013, p. 10.

This redesign effort proved controversial.[27] Most importantly, professional staff members on the House Permanent Select Committee on Intelligence were concerned that we were creating an intelligence capability. They were well aware that we had been supporting the CIA on a near-continuous basis since September 11 but did not believe that USASOC should be creating special units dedicated to what they saw as a CIA mission. However, we received strong support from the SOCOM commander, the Under Secretary of Defense for Intelligence, and the Assistant Secretary of Defense for Special Operations/Low-Intensity Conflict. All understood the importance of the mission and the toll that our support to the CIA and other agencies had taken on the teams.[28]

Although our efforts to establish and develop an unconventional warfare-focused battalion were the most controversial, the most significant aspect of the USASOC redesign was the establishment of a deployable, division-equivalent, irregular warfare-focused warfighting headquarters. Irregular warfare campaigns in which special operations is the primary defeat mechanism require a campaign-level headquarters to orchestrate the application of both surgical strike and special warfare. Yet these campaigns were always led by conventional formations, typically an Army corps or Marine expeditionary force. This new capability would partially fill that gap, although it would still lack a three-star commander and the higher standing rank of a three-star corps or Marine expeditionary force staff.

We transformed U.S. Army Special Forces Command from its previous role as the force provider for U.S. Special Forces to a deployable headquarters, consolidating all of USASOC's special warfare capabilities—Special Forces, psychological operations, civil affairs—and USASOC's 528th Sustainment Brigade into a single division-

[27] During this period, as had been the case during my time with 10th Special Forces, I was frequently asked: "Who told you to do this?" I enjoyed being able to honestly say, "Title 10, which states that USASOC's mission includes having forces trained and ready to conduct unconventional warfare."

[28] Both the Under Secretary of Defense for Intelligence and the Assistant Secretary of Defense for Special Operations/Low-Intensity Conflict were former Special Forces operators who had risen to the senior policy ranks, which was very fortunate for us.

sized headquarters. We designated this command as 1st Special Forces Command (Airborne) to align it with equivalent division capabilities in armor, cavalry, and infantry. Old Ironsides (1st Armored Division), First Team (1st Cavalry Division), and the Big Red One (1st Infantry Division) now had a Special Forces equivalent.[29]

The establishment of 1st Special Forces Command meant that the Department of Defense now had a deployable, combat capable Army headquarters staffed with men and women trained in irregular warfare. With the ability to leverage the deep expertise and knowledge of all of USASOC's special warfare capabilities, this new organization provided a powerful new capability in missions requiring an understanding of local conditions and indigenous players.[30]

This new command did not initially have the force structure to function as a campaign-level headquarters, similar to how the Marine expeditionary forces or Army corps performed for conventional operations and the Joint Special Operations Command provided for commando operations. However, my successor, Ken Tovo, would add approximately 250 billets, rounding out the necessary skills and depth to make the command much more like a division headquarters—and arguably a special operations campaign-level headquarters.[31]

We gave ourselves three years to prepare 1st Special Forces Command (Airborne) to deploy, but the situation in Syria would not give us the time we had hoped for. Just four months after its establishment,[32] the command was designated as the special operations headquarters for operations in Iraq and Syria, where it would remain deployed as Special Operations Joint Task Force–Operation Inherent Resolve for

---

[29] The new command was called 1st Special Warfare Command during the concept development phase to stay consistent with the recent doctrinal changes but was later designated 1st Special Forces Command (Airborne) to remain consistent with the Army's naming convention for its divisions.

[30] The mission statement of 1st Special Forces Command indicates that it "organizes, trains, equips, validates, and deploys regional experts in support of Theater Special Operations Commands, Joint Force Commanders, U.S. Ambassadors, and other government agencies as directed" (Cleveland, Linder, and Dempsey, 2016).

[31] Personal communication with Lieutenant General (retired) Ken Tovo.

[32] The command was provisionally active on October 2014 (USASOC, 2015, p. 9).

nearly two years. By all measures, members of the command acquitted themselves extremely well in combat, conducting near-classic foreign internal defense and unconventional warfare missions, with their efforts proving critical in the "liberation of Tikrit, Haditha, Ramadi, Fallujah and many other cities and areas in Iraq."[33]

The last major set of reforms was designed to reposition USASOC to become a thought leader for irregular warfare. This included some efforts that were strictly research-focused. Principal among these was continuing support and promotion of the Assessing Revolution and Insurgent Strategies program, a Johns Hopkins University research program into the science of resistance and revolution. This program, which was picked up from a similar program that operated during the 1950s and 1960s, produced a variety of publicly available research that I anticipate will be a cornerstone of any future irregular warfare canon, including case studies of insurgencies, resistance movements, and revolutions and several irregular warfare handbooks.[34] But we also established a new staff section (the USASOC G-9) that would be responsible for operational concept integration of special warfare and surgical strike, the two core competencies of Army special operations that we delineated in Army Doctrine Publication 3-05, and overseeing USASOC's Campaign of Learning.[35]

---

[33] USASOC, "ARSOF 2035," special issue, *Special Warfare*, 2017, p. 14.

[34] USASOC's Assessing Revolution and Insurgent Strategies program, which was initiated and managed by Chief Warrant Office (retired) Paul Tompkins in the USASOC G3X, was the continuation of a project started by the Special Operations Research Office in the 1950–1960s (Joy Rohde, *Armed with Expertise: The Militarization of American Social Research During the Cold War*, Ithaca, N.Y.: Cornell University Press, 2013). The case studies—nearly 50 in all—provide in-depth research on resistance, human factors in insurgencies, undergrounds, legal implications of the status of persons in resistances, radicalization, narratives, cyberspace, and resistance (USASOC, "Assessing Revolution and Insurgent Strategies [ARIS] Studies," webpage, undated-b).

[35] The operational concept integration had previously been in the John F. Kennedy Special Warfare Center and School, but we determined that this function would be best done at USASOC—the Army major command level—given the importance of integrating special warfare with surgical strike, aviation, and support.

The Campaign of Learning, which was borrowed from an analogous Army-wide effort,[36] got its name from a broad set of activities that USASOC would undertake to engage with others who had a role in irregular warfare or other special operations activities.[37] The most visible of these efforts was the Silent Quest future exercise, which was designed to complement the Army's Unified Quest exercise and validate special operations and irregular warfare concepts. The robust participation of conventional forces and interagency partners (e.g., the Department of State, USAID) in this exercise allowed us to test both hybrid (blending special operations and conventional) and whole-of-government concepts in a simulated irregular warfare campaign.

We also assigned civil affairs officers and psychological officers to SOCOM's offices in Washington, D.C., to establish closer relations with the Department of State's Conflict Stabilization Office, USAID, the United States Institute of Peace, and other agencies. And we similarly invested in a variety of partnerships with academia, continuing to work with the University of North Carolina system to provide academic opportunities to special operations soldiers; developing the Kennan-Donovan discussion series with Georgetown University, in which USASOC leaders and staff discussed the latest issues in the special operations community with noted (and typically young) bloggers, authors, and security sector professionals;[38] and even deploying Spe-

---

[36] This effort was led by TRADOC.

[37] USASOC's Campaign of Learning initially met resistance from TRADOC. But we quickly proved the value of our complementary effort with the Silent Quest future exercise, and the inputs from this exercise were fed into TRADOC's Unified Quest exercises to the benefit of the whole Army, as we attempted to soften the special operations–conventional divide.

[38] The Kennan-Donovan Initiative (KDI) was established at Georgetown University in conjunction with a Security Studies Program class called "Unconventional Warfare and Special Operations for Policy Makers and Strategists." The KDI convened individuals to address a major need of our country: to develop new ideas, practices, and technology in U.S. national security, with an emphasis on the advancement of special operations capabilities. The purpose was to provide a focused forum for practitioners, scholars, policymakers, congressional staff, and strategists to meet on a regular basis, over a sustained period, to address specific national security problems.

cial Forces soldiers alongside MBA-program graduates as a partnership with Notre Dame University's Business on the Front Lines initiative.[39]

The Beginnings of an American Way of Irregular War: Department of Defense–Wide Efforts to Enhance America's Irregular Warfare Capability

Although I was convinced that USASOC needed to be the proponent for irregular warfare, I recognized that we could not and should not do this alone. I was extremely fortunate that both the Army chief of staff (General Ray Odierno) and the SOCOM commander (Admiral William McRaven) recognized the importance of what we were doing and the broader value of these efforts to the Department of Defense. Their advocacy proved critical in driving a number of department-wide efforts to enhance America's irregular warfare capability.

General Odierno's support for what we were doing at USASOC dated to early 2012, when I had the chance to unveil my "Cardona slide" (see Figure 11.1) to a large group of general officers at the annual Army Training and Leader Development Conference. What I was suggesting—that there was a set of military operations that happened in the human domain for which USASOC ought to be the proponent—was almost certainly heretical to many in the room. So, I was certain that the silence from the room was either collective indignation or disbelief. But then a voice spoke up, saying: "I like this slide, it explains a lot." The speaker was General Odierno, who would become perhaps the most important advocate for the concept of the human domain and the development of Army capabilities appropriate for operations in this domain.

Admiral McRaven would similarly prove a powerful advocate for our efforts to strengthen America's irregular warfare capability. Although he had previously accused me of "making up doctrine by PowerPoint," the publication of Army Doctrine Publication 3-05 demonstrated to him the seriousness of what we were doing and the care that we were taking in articulating our arguments. I think that he

---

[39] For an example of this program, see Viva Bartkus, "'Untapped Resources' for Building Security from the Ground Up," *Joint Forces Quarterly*, Vol. 93, No. 2, May 2019.

appreciated our rigor and understood the value of our efforts for the broader special operations community.

The Strategic Landpower Task Force was the most significant of the Department of Defense–wide efforts to attempt to enhance America's irregular warfare capability that General Odierno and Admiral McRaven championed. The concept emerged from a conversation that I had with retired General Gordon Sullivan, a former Army chief of staff, about the importance of the interdependence of conventional and special operations forces in the near-continuous operations being conducted in certain critical parts of the world. The idea was that the Army should more readily weave the Army special operations narrative into its own, emphasizing that these operations were being conducted simultaneously in the land and human domains. General Sullivan declared that I was describing *strategic landpower*, a concept that he used as the Army chief of staff to make the case for a larger Army in the wake of the First Gulf War.

The central tenet in the renewed concept for strategic landpower was that advances in technology, which had made the world increasingly interconnected, meant that some nation-states (particularly those in the Middle East with strategic reserves of oil and facing increasing military threats and instability) had become part of the global commons. Protecting the global commons, which the Department of Defense defines to be the "areas of air, sea, space, and cyberspace that belong to no one state,"[40] has long been seen as a strategic priority for the United States, as maintaining freedom of movement in the air, sea, space, and (more recently) cyberspace is critical for the U.S. economy and overall national power.[41] My contention was that technology now allowed conflict, instability, and extremism to spread across countries more rapidly than ever before—and in ways that threatened the United States' and its allies' (1) access to critical resources and (2) ability to project power across the globe. In my view, the consequence was that U.S. land forces—the Army, the Marine Corps, and SOCOM—could

---

[40] Michael E. Hutchens, William D. Dries, Jason C. Perdew, Vincent D. Bryant, and Kerry E. Moores, "Joint Concept for Access and Maneuver in the Global Commons: A New Joint Operational Concept," *Joint Forces Quarterly*, Vol. 84, No. 1, 2017.

[41] Mark E. Redden and Michael P. Hughes, "Defense Planning Paradigms and the Global Commons," *Joint Forces Quarterly*, Vol. 60, No. 1, 2011.

and should make a claim that they, alongside the Air Force and Navy, had an enduring strategic mission and must maintain an equivalent level of capacity and readiness.

Building from this concept, General Odierno and Admiral McRaven established the Strategic Landpower Task Force in partnership with the Marine Corps commandant (General James Amos) in early 2013.[42] The creation of this task force reflected a growing awareness that the United States has frequently engaged "in conflict without fully considering the physical, cultural, and social environments that comprise . . . the 'human domain'" and that this oversight had contributed to our inability to achieve strategic outcomes in irregular warfare campaigns.[43] Its mission was codified in a historical document signed by General Odierno, General Amos, and Admiral McRaven, titled *Strategic Landpower: Winning the Clash of Wills*, that noted that the "significance of the 'human domain' in future conflict is growing, not diminishing" and called for the study of the "joint application of military power at the convergence of the land, cyber and 'human domains'" for these conflicts.[44]

The task force proved short-lived,[45] but it made two major contributions in its three-year lifetime.[46] The first was the development

---

[42] The inclusion of General Amos reflected the reality that the Army, Marine Corps, and SOCOM were together responsible for the vast majority of operations in the land domain in which "human outcomes . . . are a prerequisite for achieving national objectives (Odierno, Amos, and McRaven, 2013).

[43] Odierno, Amos, and McRaven, 2013.

[44] Odierno, Amos, and McRaven, 2013.

[45] The task force faced resistance from two fronts. The first was TRADOC, which would ultimately take control of the Army's portion despite the fact that it had been created to fill a gap left by TRADOC. From the onset, the Marine Corps strongly resisted the formalization of the idea of the human domain, which I suspect reflected their concern that a new domain would require dedicated resources (both talent and treasure) to develop needed doctrine, organization, training, materiel, and leaders. This concern was not without merit, as the top-down introduction of the cyber domain—which went from idea to four-star headquarters in five years—created significant requirements for the services, which affected the relatively small Marine Corps perhaps most of all.

[46] Both these contributions were thanks to Colonel Scott Kendrick who—through hard work, tenacity, and determination—convinced both TRADOC and joint doctrine bureaucracies of the importance of engaging with the founding principles of the task force. The

of the *Joint Concept for Integrated Campaigning*, which formalized the importance of campaigning well in advance of whatever culminating military event might happen.[47] It shifted military planning away from the rigid five-phase model, emphasizing that campaigning was an ongoing activity that happened before and after the military became actively engaged, and highlighted the critical roles for the interagency and other partners in early (and later) stages of campaigns. This effort, I think for the first time, really got the joint community to think about planning in a much more holistic and long-term way, which had been a critical weakness in our approach to irregular warfare.

The second was the development of the *Joint Concept for Human Aspects of Military Operations*.[48] This was a compromise document that reflected a recognition within the Department of Defense that there was a need for a human domain but that the services and the Pentagon were not prepared to establish yet another new domain (as cyber was formalized as a domain only in 2010).[49] The document did recognize that the United States needed a different set of concepts for the human equation, which was certainly needed. But it did not go nearly far enough in being able to drive the concept development, structural changes, and reform of military education needed for the United States to be more effective in population-centric problems.

The creation of the Institute for Military Support to Governance was another effective, if lesser-known, USASOC initiative that resulted

---

task force, which was formally chartered in January 2013 (U.S. Army, "Strategic Landpower Forum," August 4, 2014), was very active (e.g., developing research papers, supporting research fellows) during its early years. However, although the task force's final research paper was published in summer 2017 (Scott Kendrick, "Review of *Force Without War: U. S. Armed Forces as a Political Instrument and Implications* for Future Doctrine and Education," white paper, Strategic Landpower Task Force, July 31, 2017), the task force was effectively shuttered by summer 2016 (Matt Cavanaugh, "Strategic Landpower Is Dead: Long Live Strategic Landpower," Modern War Institute, August 14, 2016).

[47] This joint concept took several years to fully develop and was not published until 2018 (U.S. Joint Chiefs of Staff, *Joint Concept for Integrated Campaigning*, Washington, D.C., March 16, 2018), several years after the demise of the Strategic Landpower Task Force.

[48] U.S. Joint Chiefs of Staff, *Joint Concept for Human Aspects of Military Operations*, Washington, D.C., October 19, 2016.

[49] Cheryl Pellerin, "Lynn: Cyberspace Is the New Domain of Warfare," U.S. Department of Defense, October 18, 2010.

from General Odierno's support.[50] The mandate of this institute was to professionalize the process of reestablishing civil governance in post-conflict areas. Established inside USASOC with the support of the chief of Army Reserves (Lieutenant General Jeff Talley), who provided the uniquely well-qualified Brigadier General Hugh Van Roosen for the mission, the institute was also tasked with guiding "the professionalization of the [civil affairs] force structure" and creating the necessary processes to identify, credential, and commission the right civilian expertise into the Army Reserves.[51]

Although the institute initially focused on ensuring that the civil affairs community was deploying individuals with the right skills for a given mission, in later years it began to explore policy options for leveraging U.S. civilian-sector capacity for enhancing postconflict governance. The idea was that we would be much better off, in principle, if we could find a way to leverage Exxon's experience in managing oil fields around Kirkuk (in advance of their return to the Iraqi government) rather than hoping the proper expertise might be found in the reserve civil affairs pool. For this we needed an organization that could interface between our civilian sector and the military and a set of policies to govern these relationships. The hope was that the institute could both develop these policies and might even someday be that organization.

Updating the Army's warfighting construct, by codifying a key aspect of America's irregular warfare capability through the introduction of a new warfighting function, was our other major initiative. This 7th warfighting function, which was first proposed by Major General Bennet Sacolick, would "institutionalize the capabilities and skills necessary to work with host nations, regional partners and indigenous

---

[50] This institute was approved by the Army chief of staff in 2012, although it would not be established until 2013.

[51] Curtis Blais, *Governance Innovation for Security and Development: Recommendations for US Army Civil Affairs 38G Civil Sector Officers*, Monterey, Calif.: Naval Postgraduate School, 2014, p. ix. As an example, this institute studied the creation of a new military occupational specialty credentialed to help in a wide variety of governing functions, which would be able to enable the reestablishment of civilian control from the local to national levels.

populations."[52] It was an effort to codify the importance of recognizing, cultivating, and shaping interdependencies with the many other actors on the irregular warfare battlefield and reflected a "realization that the human aspect of conflict was central to success in the land domain."[53]

General Odierno chose the term *engagement* to add to the existing other six warfighting functions: movement and maneuver, intelligence, fires, sustainment, mission command, and protection.[54] *Engagement* was formalized in 2014 as the "routine contact and interaction between U.S. Army forces and with unified action partners that build trust and confidence, share information, coordinate mutual activities, and maintain influence."[55] I had recommended calling it *influence and interdependence*, which I felt more accurately captured the essence of what we were proposing, but *influence* was widely seen as politically charged. USASOC's John F. Kennedy Special Warfare Center and School was tapped to become the center of excellence for this new warfighting function, and we began establishing liaison cells in each of the other Army schools.

In the end, and despite the progress that we had made, this new warfighting function was removed from Army doctrine in 2018. The reality was that the Army was, and is, resistant to the prospect of having to broaden its warfighting precepts to fully account for irregular warfare. Without the ongoing advocacy of someone as influential as General Odierno or a combatant command–level proponent such as

---

[52] Bennet Sacolick and Wayne W. Grigsby Jr., "Special Operations/Conventional Forces Interdependence: A Critical Role in Prevent, Shape, Win," *Army Magazine*, June 2012, p. 42.

[53] See, e.g., Angela O'Mahony, Thomas S. Szayna, Christopher G. Pernin, Laurinda L. Rohn, Derek Eaton, Elizabeth Bodine-Baron, Joshua Mendelsohn, Osonde A. Osoba, Sherry Oehler, Katharina Ley Best, and Leila Bighash, *The Global Landpower Network: Recommendations for Strengthening Army Engagement*, Santa Monica, Calif.: RAND Corporation, RR-1813-A, 2017.

[54] Jan Kenneth Gleiman, *Operational Art and the Clash of Organizational Cultures: Postmortem on Special Operations as a Seventh Warfighting Function*, Fort Leavenworth, Kan.: School of Advanced Military Studies, U.S. Army Command and General Staff College, 2011.

[55] TRADOC, *U.S. Army Functional Concept for Engagement*, Fort Eustis, Va., U.S. Army Training and Doctrine Command Publication 525-8-5, February 24, 2014, p. iii.

SOCOM, reforms designed to enhance the Army's (and by extension the joint force's) irregular warfare capability are at great risk of succumbing to the Army's internal politics.[56]

An American Way of Irregular War: My Postscript

The USASOC team accomplished a great deal during my three-year tenure, laying the foundations for a reinvestment in America's irregular warfare capabilities. By claiming responsibility for irregular warfare, we were able to propose concepts and changes that had the promise of an outsized impact on fulfilling the nation's need for capacity in this form of war.

Yet, when I retired in 2015, the threat to U.S. national security from irregular war remained undiminished. The nation's two longest wars, in Afghanistan and Iraq, ground on, in one form or another; non-state actors continued to grow and threaten governments worldwide; and authoritarian regimes in China, Iran, and Russia were challenging the United States and its allies in their respective regions, largely using irregular or hybrid conflicts to challenge the United States and achieve their objectives. Conventional war remained an over-the-horizon existential threat to U.S. national security, but one in which the security sector was overly focused, leaving America most vulnerable to what remains the most prevalent form of conflict.

Early in my retirement, I would conclude that more-fundamental changes are required if the United States hopes to adequately defend its interests in the contemporary, irregular warfare–laced security environment. The United States must acknowledge that it must mature an American way of irregular war to deliver a world-class, tactical through strategic, American irregular warfare capability. To be sure, such structures must complement a robust and ready traditional war arsenal, not replace it. But a mature American way of irregular war would give

---

[56] One of my colleagues aptly described this as "TRADOC bowing to its own internal racoons, who were fighting a rear-guard action against the formal introduction of population-centric warfare unto U.S. Army doctrine."

America an improved ability to design and execute irregular warfare campaigns and support the development of better-informed policies. Ultimately, this would result in wiser presidential decisions about when and how to confront irregular warfare threats and wage surrogate, proxy, or limited war.

CHAPTER TWELVE

Key Observations

As I look back on the population-centric campaigns in which I was involved, I am struck by their similarity to our failures in Vietnam. None was nearly as costly as Vietnam, where more than 58,000 Americans lost their lives. But in each we never lost a battle yet still failed to achieve victory at the strategic level.

Our difficulties in Afghanistan and Iraq, eerily reminiscent of the U.S. experience in Vietnam, are perhaps the most pronounced examples. But the same was true of Bolivia, which was the U.S. frontier in the war on drugs and my first foray into irregular warfare, as the Bolivian government eventually expelled the United States and legalized coca after a failed 20-year (and multibillion-dollar) campaign. In El Salvador, we were happy to call the negotiated settlement with the Soviet-supported insurgent a success, although it eventually would become a basket case and the home of the terrifying Mara Salvatrucha (MS-13) crime syndicate. Even the peace that we brokered in Bosnia now threatens to unravel, despite our enduring military presence.[1]

In this chapter I provide what I see as the six key observations from my 37 years in uniform as a student and practitioner of irregular warfare. Although these six observations are colored by the many times that I spent with U.S. Special Forces, I conclude each discussion by providing the perspectives of the researchers and practitioners consulted during this project. I hope that these observations may offer a

---

[1] Mark MacKinnon, "In Bosnia-Herzegovina, Fears Are Growing That the Carefully Constructed Peace Is Starting to Unravel," *Globe and Mail*, May 24, 2018; Srecko Latal, "Bosnia's Dayton Deal Is Dead—So What Now?" *Balkan Insight*, December 2, 2019.

new perspective on why we have struggled to achieve strategic victory in population-centric contests.

Observation 1: U.S. Tactical-Level Formations Have Performed Admirably in Irregular Warfare Campaigns

Throughout my career, our tactical formations succeeded in every irregular warfare mission that I saw put before them. The intelligence, creativity, flexibility, and professionalism of operators across our national security enterprise are the reason we never lost a battle. This was certainly true of Special Forces, the only U.S. force dedicated to this form of warfare and with whom I spent most my career. But it was also true of the multitude of soldiers, marines, Navy SEALs, and civilian special operations whom I would serve with.

My perspective is certainly colored by my own personal experience, but I saw our U.S. Special Forces thrive in these missions time and time again. This is perhaps unsurprising, as our Special Forces are selected and trained for precisely these types of conflicts, enabled by their language ability, understanding of the roots of resistance and how to counter or use it, and tradecraft that allows them to work in hostile or even denied areas. They are also prepared for the persistence and sacrifice that this mission sometimes requires.

A unique capability that our Special Forces brought to bear in these missions was the ability to build partner forces from existing indigenous capabilities in almost any context imaginable—whether that be in the drug trafficker safe havens in the jungles of Bolivia, among the crime syndicates and labor unions in the suburbs of Sarajevo, with resistance forces in Afghanistan (Northern Alliance) and Iraq (Peshmerga), or in the wild west that is Pakistan's Federally Administered Tribal Areas. In some cases, the relationships were short term, but in others the Special Forces would spend more than a decade building, training, and fighting alongside these partners, as was the case with the Iraqi Counter Terrorism Service and Afghan commandos. These partners, when provided appropriate and tailored U.S. technology, firepower, and knowledge, have proved to be a critical tool in furthering U.S. objectives.

Increasingly, units outside our Special Forces teams have been integrated into efforts to leverage indigenous mass, and they too have succeeded in these roles. This includes both Special Forces parallel capabilities in the Marine Corps and the Navy, the Marine Corps special operations teams, and Navy SEALs, who played key roles in the irregular warfare missions in Afghanistan and the Philippines (and elsewhere). I have been equally impressed by the critical roles that conventional forces have played in U.S. successes in this form of warfare. The efficacy of the 26th Marine Expeditionary Unit in quickly going from theater reserve to supporting the management of Mosul after the initial invasion of Iraq in 2003 and the indispensable role that conventional "uplift" played in enabling our special operations mission in Afghanistan are only the two examples I remember the most. It must be said that conventional units have frequently had more difficulty in creating enduring combat capable indigenous units, but there have been advances. The Security Force Assistance Brigades, the Asymmetric Warfare Group, and the Marine Corps Irregular Warfare Regiment are positive steps.

Partnerships with U.S. civilian special operations have also frequently been central in tactical-level successes in which I have been involved. In Bolivia and Colombia, it simply would have been impossible to build credible partner-nation capabilities, which help create the domestic will needed to confront such mutual problems, without our in-country partnerships with the Department of State and DEA. Similarly, the relationship between military special operations and the CIA, however unorthodox at the time, contributed to our rapid success against the Taliban in Afghanistan in 2001. An analogous partnership, involving the CIA, conventional military forces, and military special operations enabled our success in toppling Saddam Hussein just two years later.

There was broad agreement among the researchers and practitioners consulted during this study that the tactical-level irregular warfare capabilities within the Department of Defense are largely sufficient. As examples, one respondent concluded succinctly that "we are not lacking in technical or tactical capabilities," and another indicated that the United States "possesses discrete capabilities at the tactical level (e.g., FID [foreign internal defense], UW [unconventional warfare], MISO [military information support operations]) relevant to competing in this domain of warfare."[2] Several of these respondents emphasized U.S. special operations capabilities in their comments, with one respondent indicating that the "U.S. military, specifically [special operations], have the capabilities to contribute to an effective whole-of-government solution to achieve success in competition short of traditional armed conflict."[3]

However, several respondents believed that the capabilities of other elements of America's irregular warfare capabilities, the Department of State and USAID, in particular, were less developed. One respondent concluded that the "non–Department of Defense partners are weak and growing weaker," and another highlighted the human resources challenge that USAID has faced, as it has consistently been unable "to identify, recruit, deploy, and retain the appropriate personnel."[4]

Observation 2: Irregular Warfare Missions Require Irregular Warfare Campaigns

This observation is almost a tautology, but the United States has been successful in irregular war when it develops campaigns that are appropriate for the context and the type of adversary that we face. Effective irregular warfare campaigns reflect a deep understanding of the local context and adversary, set planning horizons that are appropriate for the challenge, and rely on local solutions for local problems. These

---

[2]   The first quote is from the project workshop's participant 18, and second is from participant 2.

[3]   The project workshop's participant 11.

[4]   First quote is from the project workshop's participant 6, and second is from participant 19.

facts are widely understood,[5] but the United States has struggled to develop military campaigns appropriate for irregular war.[6]

The United States has excelled in three types of irregular warfare campaigns. The first are campaigns that were purpose-designed to be long duration and whole of government and rely on indigenous approaches. Unfortunately, I can think of only one U.S.-led campaign that really fits this definition: Plan Colombia. This U.S. Embassy–led program supported the Colombian government in fighting drug cartels and combating the domestic insurgency during 2000–2015. Plan Colombia was deliberately long duration; emphasized partnering by civilian law enforcement and military forces with Colombian counterparts, allowing the Colombians to be the counterinsurgents; and leveraged a whole-of-government approach that blended U.S. development assistance, military assistance, and diplomatic support. It helped also that the program was born bipartisan and with a rare and powerful consensus between the Executive and Legislative Branches. In many ways, Plan Colombia is the exception that proves the rule, as it was designed and led by the Department of State.

We have also been reasonably successful in small-footprint operations. Most notable among these, perhaps, were the campaigns in Bolivia, El Salvador, and the Philippines in which U.S. Special Forces, selected conventional capabilities, and representatives from the U.S. intelligence community advised, enabled, and supported indigenous partners. The small footprint of these operations allowed us to both sustain long-duration missions and ensure that we did not relieve the host nation from shouldering the burden of its fight, which would become a major failure point (in my view) in the "big wars" in Afghanistan, Iraq, and Vietnam. These efforts might also have, inadvertently, benefited from a sort of benign neglect in that political considerations and the lingering "Vietnam syndrome" prevented them from becom-

---

[5]   U.S. Department of Defense, 2007.

[6]   As an example, see Ben Connable, *Redesigning Strategy for Irregular War: Improving Strategic Design for Planners and Policymakers to Help Defeat Groups Like the Islamic State*, Santa Monica, Calif.: RAND Corporation, WR-1172-OSD, 2017.

ing a U.S. conventional theater of war.[7] The major challenge that we face in these small-footprint operations, which I see now as I look back on what has become of Bolivia and El Salvador, is that we struggle to convert tactical achievements into strategic successes.

The third are campaigns that are short duration and with limited and well-defined strategic objectives. Our opening gambits in Afghanistan in 2001 and Iraq in 2003, the month-long mission in Panama to remove General Manuel Noriega, the "100-hour war" that booted Iraq out of Kuwait, and the even shorter four-day operation to remove a communist from power in Grenada are demonstrative of this success.

However successful, these limited-duration campaigns also illustrate a major weakness in how the United States plans for irregular warfare—specifically, that we have little idea about what to do after the initial phase of the invasion. Although Panama, with its unique geography and history, has succeeded, both Afghanistan and Iraq have proven problematic, almost to the point of failure.

The success that we found in the "big fights" in Afghanistan and Iraq relied on our ability to effectively leverage indigenous mass, made possible through close relationships built on mutual trust and respect between our American warriors and their host-nation counterparts. This came from shared sacrifice over years of training and fighting together, perhaps best illustrated by the Iraqi Counter Terrorism Service, which led the defense of Baghdad in 2014 and was the decisive force in the retaking of Mosul and defeat of the Islamic State, as well as the Afghan commandos, who have become the primary maneuver capability of the Afghan National Army. But these were partnerships that came much later and through the intelligence, creativity, flexibility, and professionalism of our troops on the battlefield rather than good planning.

The perspectives from my interviews and the workshop align closely with this conclusion. Many of the representatives from the Department of Defense spoke specifically of the lack of appropriate campaigns for irregular warfare, with one exemplar respondent concluding that "the United States is greatly disadvantaged when trying

---

[7]   Rothstein, 2007.

to deter contemporary Russian aggression by our inability to conduct an effective irregular warfare campaign."[8] Others spoke more generally of the characteristics of a successful irregular warfare campaign that I highlighted above. In particular, individuals noted the difficulty that the United States faces in transitioning after initial combat operations and in designing and following through on campaigns that are sufficiently long term, as the "process takes much longer than allowed by our leadership," and the United States "disengages too quickly, both militarily and financially."[9]

Observation 3: The U.S. Military Is Not Well Organized for Irregular Warfare Campaigns

The United States has never maintained a standing capability for designing, conducting, or overseeing irregular warfare campaigns. Instead, the prevailing assumption has been that the headquarters and formations built for traditional war can be made to work in irregular war. We saw this in Vietnam, when the United States poured conventional formations into the country to conduct counterinsurgency operations, after we had first supported a democratically elected government and later fomented a coup. In Bosnia, the U.S. military deployed conventional forces in something akin to postwar occupation duty, with that "presence" giving time for diplomacy, economic initiatives, and other elements of U.S. and other Western soft power to do their work. Afghanistan was similarly transitioned to conventional forces after our defeat of the Taliban in 2001.

This reliance on the same theater- and campaign-level headquarters for conventional and irregular warfare has had clear consequences for our efficacy in irregular warfare. For one, these headquarters are not prepared to develop or execute irregular warfare campaigns or to defend against those of our adversaries. They simply lack the expertise and necessary experience in irregular warfare, as well as an appre-

---

[8]   The project workshop's participant 4.

[9]   First quote is from the project workshop's participant 19, and second is from participant 12.

ciation of the important role of the host country in such a campaign. Resources—money, equipment, and personnel—are frequently viewed as a substitute for time, or a way for overcoming host-nation resistance and erode the legitimacy of our efforts and our host-nation partners.

As a result, I watched these conventional formations struggle to translate tactical successes into something more enduring in nearly every population-centric campaign in which I was involved. They brought tremendous capabilities to the fight, but their potency in orchestrating mass, maneuver, and fires was often of little direct relevance to the causes of the insurgencies and resistance movements they were fighting. It was perhaps unsurprising, then, that the Seventh Army struggled to understand the complexities of Bosnia, XVIII Airborne Corps was unprepared to build an Afghan National Army in 2002, and I Marine Expeditionary Force and V Corps struggled to rebuild the Iraqi Army.

There is no question that these headquarters and their staffs improved dramatically during their deployments, improvising to create new processes and structures in an effort to be effective. The multitude of new structures that would make an appearance on the battlefield are a testimony to the ingenuity and dedication of our military professionals—Cultural Support Teams, Embedded Training Teams, Female Engagement Teams, Human Terrain Teams, Mobile Training Teams, Provincial Reconstruction Teams, Special Police Transition Teams, and many more. These structures were often quite effective in providing the conventional headquarters a mechanism to work with locals, foreign units, other U.S. government agencies, and international civilians. But they were ad hoc and took time to develop and mature, and each arriving headquarters and staff would use these capabilities differently.

Our conventional formations were not, are not, and should not be prepared for this form of warfare. Following Vietnam, we reoriented our conventional forces toward traditional war, and the wisdom of that decision was undeniably validated in Operation Desert Storm. Asking these same formations, optimized now for AirLand Battle, to play the role of occupier while conducting counterinsurgency operations, training and equipping a partner force, and working with other agencies to bring democracy and good governance is, in my view, folly. In addition to being ineffective, it threatens to unravel our conventional dominance, which is critical given the role of this conventional might as a deterrent to our Great Power adversaries.

The question remains, however, of how the United States can leverage indigenous mass at scale. The Department of Defense, despite the prominence of these population-centric conflicts, has remain focused on the conventional fight, and we have yet to begin to develop the structures to defend against or conduct irregular warfare at the campaign or operational level.

Admittedly, the range of experts whom I consulted for this project might have been somewhat biased in this regard, but there was broad agreement with this conclusion. One respondent summarized the issue succinctly and plainly: "I believe that we can conduct small-scale operations, but we are incapable of conducting large-scale irregular warfare operations."[10] One senior-level interlocutor suggested, matching my own experience, that the challenge may be in large part because SOCOM's "organizational structure perpetuates a focus on direct action at the expense of indirect approaches."[11] That said, others disagreed with my assessment that conventional formations were not adaptable to irregular warfare, with one individual suggesting that the Department of Defense ought to "rework the mind-set and capabilities" of these formations to prepare them for irregular warfare.[12]

Observation 4: The United States Lacks the Concepts, Doctrine, and Canon Necessary to Be Effective in Population-Centric Conflicts

Today, we still have neither the concepts nor doctrine to achieve our stated goals in population-centric conflicts. During my career, I was part of several efforts designed to fill this gap and observed several

---

[10] The project workshop's participant 4.

[11] The project workshop's participant 11.

[12] The project workshop's participant 5.

others, but none produced anything close to the mature thinking that the U.S. military has developed for conventional war. In addition, we have yet to develop canon—specifically, what constitutes good or bad thinking, practices, and policy approaches—for this form of warfare.

Population-centric conflicts cannot be fought with military concepts and doctrine designed for the physics of conventional war and instead require approaches that blend anthropology, economics, history, and sociology. The resolution of population-centric wars is necessarily indigenous in form, requiring solutions that are appropriate for the local and regional context. These indigenous-driven campaigns typically take longer, are less efficient, and sometimes require compromising on selected values until the situation allows for the offending activities to be addressed without having an impact on the effectiveness of the military campaign. Further, these conflicts are usually aggravated by the forces, money, and culture of outsiders.

This lack of concepts, doctrine, and canon (as I call it) is perhaps a bit perplexing to some who, like me, have been avid readers of what has been a truly astounding quantity of writing on irregular warfare. The challenge is not the quantity of writing on this topic, by academics, practitioners, and pundits. The challenge is instead that there has never been a deliberate, widely accepted process to codify the lessons from the United States' misadventures with irregular warfare and to describe how we ought to organize ourselves and how we ought to plan for this form of warfare.

We have retained (and continue to develop) substantial thinking about how to operate in these conflicts at the tactical level, but efforts to develop the type of thinking necessary to achieve strategic success in irregular warfare have been less successful. The much-ballyhooed collaboration of the Army and Marine Corps in producing and then revising Field Manual 3-24, which was originally *Counterinsurgency* and then became *Insurgencies and Countering Insurgencies*,[13] is a case in point: The manual assumes that U.S. forces are occupiers and therefore guarantors of security for a foreign population. It thus does well

---

[13] Field Manual 3-24 and Marine Corps Warfighting Publication 3-33.5, 2006; Field Manual 3-24 and Marine Corps Warfighting Publication 3-33.5, 2014.

enough in describing how conventional forces should approach counterinsurgency but remains of limited utility in campaigns not built on occupation and nation building.

Two related efforts that I was involved in during my career have similarly had, at best, a modest impact on America's thinking in this domain. The drafting of Army Doctrine Publication 3-05, *Special Operations*, was a step in the right direction in terms of how the United States should think about the role of Army special operations in warfare—and in irregular warfare in particular. But although it advanced our thinking on *why* the United States should sustain an irregular warfare capability, it did little to answer the question of *how* the United States should respond.[14]

The most promising initiative that sought to explore the question of "how" the United States should respond was the penning of the unprecedented *Strategic Landpower: Winning the Clash of Wills*. In this document, the U.S. Army chief of staff, U.S. Marine Corps commandant, and SOCOM commander jointly acknowledged the growing contemporary threat from irregular adversaries and the need to develop proper responses. Each leader faced tremendous headwinds from the establishment below the leadership, and the progress they made was largely gone by the time that their replacements arrived.

The problem, in my view, is that there is no proponent within the U.S. government to drive investment in the fundamentals necessary for maturing our irregular warfare capability. The United States is to fight wars in physical domains—air, cyber, land, sea, and space—with each service developing concepts and doctrine to support maneuver warfare in their dominant domain. In the words of others, population-centric conflicts are "a profoundly human activity," and the focus on physical domains has led us to "overlook, and underinvest in, the more important aspects of war and warfare—those best defined as human."[15] The

---

[14] Army Doctrine Publication 3-05, 2012. See Kiras, 2015, for a relevant discussion in the context of the debate on whether there is a need for a generalized theory for special operations.

[15] Michael C. Davies and Frank Hoffman, "Joint Force 2020 and the Human Domain: Time for a New Conceptual Framework?" *Small Wars Journal*, June 2013.

result is that we do not have effective concepts or the right tools for succeeding, let alone dominating, in these population-centric conflicts.[16]

In other words, we lack an irregular warfare-focused superstructure above the tactical level that can consistently push for the needed development of concepts, doctrine, and canon. There is no dedicated three-star, irregular warfare–focused headquarters (equivalent to an Army corps or Marine expeditionary force) that can direct and arrange activities and resources in the form of plans and operations. And there is no capability at the institutional level, where war is studied, concepts are developed, and the education of its practitioners takes place. SOCOM was established in part to fill this gap, but today, nearly 30 years after the Senate Armed Services Committee censored SOCOM for not giving sufficient priority to the low-intensity conflict components of its portfolio,[17] many simply do not see the development of America's irregular warfare capabilities as a priority.

Among the six observations, this is the one observation in which the support among the consulted experts was somewhat more mixed. Although few respondents spoke specifically of doctrine or canon, there was some significant discussion about whether we had sufficient understanding to be effective in population-centric conflicts. Representatives from the U.S. intelligence community thought that "understanding the nature of the conflict was not usually the problem," and a policy expert concluded that understanding is not typically a concern "in areas where our military has extensive experience on the ground."[18] Another concluded that the challenge is typically not a lack of understanding within the U.S. government but rather that decisionmakers often do not take seriously dissenting perspectives from other governmental agencies: "Senior leaders get plenty of input from the expert level at State, but we have seen many examples of mistakes because they did not read it."[19]

---

[16] Odierno, Amos, and McRaven, 2013.

[17] Adams, 1998, pp. 205–206.

[18] The first quote is from the project workshop's participant 10, and second is from participant 18.

[19] The project workshop's participant 14.

Observation 5: There Is Insufficient Professional Military Education for Irregular Warfare

The U.S. military does not provide the officers responsible for irregular war with the education necessary to be effective. This affects our officers at the vanguard of our tactical-level irregular warfare efforts (e.g., Special Forces), who receive largely the same education as their conventional counterparts,[20] although these officers can and do try to fill in their gaps in knowledge through either self-study or experiential learning.[21] However, this has a significant effect on other officers, both special operators with a background in more-traditional warfare (e.g., Rangers) and others, who are often ill-equipped to lead irregular warfare–focused formations at the tactical level.

The Department of Defense, the services, and SOCOM have all acknowledged this gap and highlighted the importance of including some irregular warfare into professional military education. As

---

[20] There is widespread agreement that the professional military education offered to special operations officers is problematic (Bryan Cannady, "Irregular Warfare: Special Operations Joint Professional Military Education Transformation," thesis, Fort Leavenworth, Kan.: U.S. Army Command and General Staff College, 2008; Eli G. Mitchell, *Special Operations Professional Military Education for Field Grade Officers*, Maxwell Air Force Base, Ala.: Air Command and Staff College, 2015). Although these officers are likely to be practitioners of irregular warfare at some point in their careers, their training is largely the same as that of conventional officers, as there is not a "consolidated SOF educational trajectory" (Barak A. Salmoni, Jessica Hart, Renny McPherson, and Aidan Kirby Win, "Growing Strategic Leaders for Future Conflict," *Parameters*, September–October 2018, p. 84). The one exception to this is the Naval Postgraduate School, which has had an outsized impact on the education of practitioners of irregular warfare. However, it faces yearly assaults from the Army and Navy to justify its Defense Analysis Department's special operations curriculum and the number of Army special operations officers who attend. Further, far from being a part of an acknowledged pipeline for U.S. irregular warfare military experts, a relatively small percentage of the nation's special operations professionals attend, and indeed the Army sees it as a drain of its special operations officers who should be at Fort Leavenworth attending the Command and General Staff College with their conventional counterparts. There is no recognition that officers whose chosen profession is irregular warfare require time to become field-grade thinkers and planners in that distinct form of war (Cannady, 2008; Mitchell, 2015; Salmoni et al., 2018, p. 84).

[21] E.g., Philip A. Buswell, "Keeping Special Forces Special: Regional Proficiency in Special Forces," thesis, Monterey, Calif.: Naval Postgraduate School, 2011.

an example, the Joint Staff's education policy indicates that irregular warfare "is as strategically important as traditional warfare."[22] And the services have integrated irregular warfare curriculum into their core coursework,[23] created electives focused on irregular warfare,[24] and established specialized schools focused on the study of irregular warfare.[25]

Despite these efforts, and some notable successes, irregular warfare education is still inadequate.[26] This is, in part, a consequence of the lack of concepts and doctrine for irregular warfare. However, professional military education institutions, by necessity, continue to favor the proper application of service forces in traditional warfare, essentially the requirements of dominating the domain for which the service was built. Land for the Army, sea for the Navy, air for the Air Force, space for the Space Force, and even cyber has linked service forces. These same schools do not see it as their mission to educate professional practitioners of population-centric and irregular wars.[27]

---

[22] U.S. Joint Chiefs of Staff, *Officer Professional Military Education Policy (OPMEP)*, Washington, D.C., September 5, 2012.

[23] U.S. House Armed Services Committee, *Another Crossroads? Professional Military Education Two Decades After the Goldwater-Nichols Act and the Skelton Panel*, Washington, D.C., 2010, p. 74.

[24] U.S. House Armed Services Committee, 2010, p. 74.

[25] One example is the Naval War College's Center on Irregular Warfare and Armed Groups.

[26] See, e.g., U.S. House Armed Services Committee, *Institutionalizing Irregular Warfare Capabilities: Hearing Before the Subcommittee on Emerging Threats and Capabilities of the Committee on Armed Services, House of Representatives*, Washington, D.C.: Government Printing Office, 2012; U.S. Air Force, *Irregular Warfare Strategy*, Washington, D.C., 2013, p. i.

[27] SOCOM has but has been prohibited by Congress from spending its funds on developing special operations–specific education because Congress believes this to be a service-level function. Admiral Eric Olson advocated—in 2009—for an "increase in SOCOM's involvement in the management of personnel," to include education. However, this faced stiff resistance from the services that believed education to be a service-level responsibility (U.S. Senate, *Department of Defense Authorization for Appropriations for Fiscal Year 2010: Hearing Before the Committee on Armed Services, United States Senate*, Part 5, Washington, D.C.: Government Printing Office, 2009). This misguided view from Congress leaves the officers whose focus is irregular warfare reliant on education developed by the services, which have a very different primary focus—hardly the hallmark of a profession. Recent legislation may

Without a specialized, and ideally standardized, professional military education mechanism to prepare our officers, the United States is not preparing the senior field-grade and general officers that it needs to conduct irregular warfare at the operational and campaign levels. This was illustrated perhaps must dramatically in the run-up to the invasion of Afghanistan in 2001, when Secretary of Defense Donald Rumsfeld famously disparaged the Department of Defense's ability to generate anything but conventional options before choosing the hybrid CIA, Special Forces, and Air Force option for the initial invasion of Afghanistan. But in the years that followed, our senior leaders have consistently demonstrated a hesitancy to apply approaches that rely on indigenous forces and have a poor appreciation of what host-nation forces can do and how quickly they can do it.

The experts I consulted spoke less of whether professional military education was sufficient and more about whether we had senior military and civilian experts who were sufficiently proficient in this form of warfare. Here again there was some disagreement among my interlocutors. Several of the experts agreed with my own assessment, with one respondent summarizing this perspective in concluding that "senior military and civilian leaders have failed to appropriately and succinctly define the threat and identify relevant U.S. government capabilities and policies to operate in the domain (as well as identify gaps)."[28] Others highlighted this as a particular challenge in the Department of Defense: "The advice given to senior policymakers by the Department

---

help address this issue, as Section 922 of the 2017 National Defense Authorization Act gave the Assistant Secretary of Defense for Special Operations/Low Intensity Conflict "authority, direction, and control of all special-operations peculiar administrative matters relating to the organization, training, and equipping of special operations forces," as well as the responsibility to assist in the "development and supervision of policy, program planning and execution, and allocation and use of resources" for irregular warfare, as well as counterterrorism and special operations-specific activities (Pub. L. 114-328, National Defense Authorization Act for Fiscal Year 2017, December 23, 2017). That said, it has been two years since this was published, and we have yet to see significant progress in this domain.

[28] The project workshop's participant 2.

of Defense is unbalanced and very much biased toward armed conflict. . . . This gap opens space for insurgencies to thrive."[29]

However, others disagreed and concluded that the military advice being provided in irregular warfare was either "sufficient" or, in one case, "superb." These individuals offered two broad alternative explanations for why poor decisionmaking often resulted. One explanation was that the interagency process was failing: "Senior military leaders can provide military advice, but those activities only serve to buy time for other interagency activities. . . . Unfortunately, those interagency activities are frequently too little and too late."[30] The other proposed explanation was that "problem lies with political guidance . . . [and] generally weak policy formation process in Washington."[31]

Observation 6: The U.S. National Security Enterprise Is Structured to Fail in Population-Centric Conflicts

The national-level bureaucracy that supports U.S. war-waging activities is inadvertently structured to fail in population-centric conflicts. Before Vietnam, people on the battlefield were largely inconsequential, and cities were to be bypassed, cordoned off, or occupied. But with the advent of television (and later the internet), traditional military instruments could no longer be used in such conflicts in all their brutality without substantial political and legal consequences.[32]

Our 70-year-old national security architecture was created to contest Great Power adversaries and not for the adversaries that we would face in the years that followed. As a result, the U.S. national secu-

---

[29] The project workshop's participant 4.

[30] The project workshop's participant 11.

[31] The project workshop's participant 13. This respondent indicated that "the commanders in the field have open channels of communication to flag officers and can provide dissenting information—the process of evaluating information from the field is also excellent."

[32] And although managed state violence still has an important role, it can inadvertently feed insurgencies if improperly applied.

rity enterprise has been dramatically reformed twice during my career, both in the wake of dramatic failure.

The first, which came in the wake of the bumbled efforts to rescue U.S. hostages from our embassy in Iran (Operation Eagle Claw), forced "jointness" on the Department of Defense and eventually compelled the formation of SOCOM. This multiyear process resulted in the most proficient counterterrorism capability in the world, a capability that the United States has depended on while bringing the fight to America's modern adversaries. The second reform was forced on the U.S. intelligence community by Congress in 2004,[33] giving the United States the political and resource assistance to organize effectively against al Qaeda.

The U.S. national security enterprise is similarly not well structured for success in irregular warfare. For one, there is no proponent for irregular warfare in the any of the services in the Department of Defense, Department of State, CIA, National Security Council, or anywhere else within the Executive Branch of the U.S. government.

But perhaps even more importantly, there is no superstructure to allow the United States to design and implement the whole-of-government, and potentially whole-of-society, solutions that are necessary for efficacy in this form of conflict.[34] Congress does not fund whole-of-government programs, nor does it have a committee structure that champions such efforts, and the executive rarely develops cross-department programs or seeks funding for them. Restrictions on working with industry and academia are even more formidable.[35]

---

[33] This was Pub. L. 108-458, Intelligence Reform and Terrorism Prevention Act of 2004, December 17, 2004, sponsored by Senators Susan Collins and Joe Lieberman.

[34] Charles T. Cleveland, Ryan Crocker, Daniel Egel, Andrew M. Liepman, and David Maxwell, *An American Way of Political Warfare: A Proposal*, Santa Monica, Calif.: RAND Corporation, PE-304, 2018. Indeed, the ongoing technological revolution will likely require this type of superstructure, if we hope to maintain a competitive edge. The advances in artificial intelligence, explosive growth in smartphone and social media use, and accelerating advancement of sensor technology provide an opportunity to detect the early beginnings of instability and unrest, offering the chance to perhaps, for pennies on the dollar, prevent conflict.

[35] For example, the legal restrictions on funding military participation in partnering efforts, such as Notre Dame's groundbreaking Business on the Frontlines program, killed a poten-

A lack of coordination across the U.S. national security enterprise was highlighted by most experts I engaged with. One respondent's perspective agreed very closely with my own, indicating that the United States "lacks a whole-of-government strategy and plan to employ all of its instruments of national power in a synergized and integrated manner."[36] Another concluded that the combatant commands "are not effectively integrated with the country teams, State Department, or USAID," and a third reported that "a lack of coordination across the enterprise undermines ability to effectively operate."[37] One respondent went so far as to conclude that "the lack of a coherent interagency process to effectively combine irregular warfare with strategic communications, diplomacy, and development" contributed to the failure of policymakers to embrace this approach to warfare and contributed to the failed decisions to "not intervene more actively in Syria to counter the Assad regime and to counter the Russian annexation of Crimea."[38]

---

tially powerful approach to conflict prevention and postconflict mitigation. Spirit of America, a remarkable example of American entrepreneurial energy and intellect that harnesses the goodwill of businesses and philanthropists to support our troops, offers another important example (USASOC, "Spirit of America," *Special Warfare*, January–June 2016, p. 45). Its efforts to turn contributions into needed supplies or services that the logistics or administration systems find hard to provide encountered similar legal challenges (Dan De Luce, "Meet the Venture Capitalist Who Launched a Kickstarter for War," *Foreign Policy*, May 31, 2017).

[36] The project workshop's participant 5.

[37] First quote is from the project workshop's participant 11, and second is from participant 2.

[38] The project workshop's participant 11.

CHAPTER THIRTEEN

Recommendations

There is no question that the United States has faced persistent difficulty in achieving strategic objectives in population-centric campaigns. And the cost of these campaigns has been very high, in both wasted life and treasure. During my career alone, the troubled campaigns in Afghanistan and Iraq, against the Islamic State, and against irregular forces in Somalia, Yemen, and Libya cost the American taxpayers trillions of dollars, and tens of thousands of Americans have been killed or wounded.[1]

It is my belief that we could have done much better in these conflicts, and will do much better in future conflicts of this ilk, if the United States reorganizes itself to take irregular warfare seriously. Yet it is unlikely that the needed reform will come from within. In the wake of Vietnam, the U.S. military deliberately turned its back on the irregular warfare, vowing never to fight that kind of war again.

There is a real risk that the "bureaucracy will do its thing" and we will again pivot away from irregular warfare.[2] The likelihood of this happening is demonstrated perhaps most clearly by the failure of the

---

[1] Congressional Research Service, "U.S. War Costs, Casualties, and Personnel Levels Since 9/11," Washington, D.C., April 18, 2019.

[2] This is a reference to Robert Komer's influential work that documented how difficult it is to get bureaucracies to "keep doing the familiar" (Robert W. Komer, *Bureaucracy Does Its Thing: Institutional Constraints on U.S.-GVN Performance in Vietnam*, Santa Monica, Calif.: RAND Corporation, R-967-ARPA, 1972, p. xii). His observations were specific to how the United States and the government of Vietnam struggled to adapt to failure during the Vietnam War, although the equivalent argument has been made for Afghanistan (Todd Greentree, "Bureaucracy Does Its Thing: US Performance and the Institutional Dimension

Strategic Landpower: Winning the Clash of Wills—if anything was to generate change, it would have been this historic document, in which Army Chief of Staff General Ray Odierno, Marine Corps Commandant General James Amos, and SOCOM Commander Admiral William McRaven came together to acknowledge that there was a problem and something ought to be done. But it has apparently already disappeared into the annals of history.[3]

In the wake of the coordination failure that led to the failed Operation Eagle Claw and the intelligence failure that led to September 11, it took action by Congress and the support of the President to drive the reforms that we needed to maintain lethality against our modern adversary. I believe that Congress and the President will need to act again if we hope to develop the American way of irregular war that we need to provide a proactive defense and the offensive potential to destabilize our Great Power adversaries.

My three recommendations for how we can do better in these conflicts are therefore drastic and detail approaches that the U.S. Congress, President, and a group of well-financed and concerned citizens could take to force the needed reforms of the U.S. national security enterprise.

My concern is that, despite our acknowledged failures, there seems to be a prevailing belief that the United States can simply retool whenever an irregular warfare capability is needed. But my experience is that our irregular warfare capability has been insufficient even when at its peak. In my view, and based on the conversations I have had while developing this study, drastic measures, such as those taken in Goldwater-Nichols, Nunn-Cohen, or Collins-Lieberman, are needed if

---

of Strategy in Afghanistan," *Journal of Strategic Studies*, Vol. 36, No. 3, 2013) and can be more generally applied to the post-Vietnam experience.

[3] Odierno, Amos, and McRaven, 2013. This document is not discoverable on any publicly available Department of Defense systems, and the Strategic Landpower Task Force has similarly disappeared.

we want the national security enterprise to be prepared for this form of warfare.[4]

These three recommendations are interdependent, with each necessary but insufficient by itself to achieve the needed changes in America's irregular warfare capabilities. Much like General Peter Schoomaker's analogy of the Rubik's Cube during my time in Bosnia, developing the American way of irregular war that we need will require a full understanding of today's challenges and the support of Congress and the President to make the changes necessary.

Recommendation to Congress: Mandate an Independent Review of U.S. Strategic Failure in Population-Centric Conflicts

For all the blood spilled and the money spent in population-centric conflicts, over the past two decades in particular but in the decades prior as well, there has been a surprising lack of introspection within the U.S. national security enterprise. There has also been surprisingly little public pressure for Congress to act. This stands in sharp contrast to congressional action in the wake of other strategic failures from our national security enterprise.

Regardless, it is past time that Congress get involved and demand a thorough review and public accounting of U.S. performance in these population-centric conflicts. To do so, Congress should empanel a bipartisan commission with the mandate of making recommendations on how the United States can improve its policies, strategies, and campaigns in these conflicts. This commission would assess U.S. performance in achieving strategic objectives in population-centric conflicts of all types, with the intent of diagnosing why the U.S. national secu-

---

4   Pub. L. 99-433, Goldwater-Nichols Department of Defense Reorganization Act of 1986, October 1, 1986; Pub. L. 99-661, 1986; Pub. L. 108-458, 2004.

rity enterprise has failed to deliver strategic success in many of these conflicts and how it can be improved.[5]

This review would examine (among other topics) the decision underlying the large-scale deployment of conventional troops, the risk and cost of the current "boom and bust" approach to irregular warfare, and the roles and interoperability of different elements of U.S. national power (e.g., Department of Defense, Department of State, CIA). In addition, this analysis should examine whether Congress and the Executive Branch are appropriately designed for success in this form of warfare. It has been my observation that irregular warfare is split among the armed services, foreign affairs, and intelligence oversight committees, and that in itself may be part of the difficulty that we have faced in achieving strategic effects in these contests.

The task force undertaking this review would study America's application of irregular warfare since (and including) World War II, assessing our own techniques, the techniques that our adversaries used against us, and whether existing concepts (such as multidomain operations) are sufficient for this form of warfare. The task force would then use this information to examine the potential roles of key types of personnel (military, foreign service, intelligence) in irregular warfare; identify missing doctrine, capabilities, and structures; evaluate existing proposals for enhancing the U.S. irregular warfare capability; explore the possibility of public-private irregular warfare capabilities; and determine organizational changes necessary for the efficacy of the United States in these contests.

The intent of this systematic review would be to produce recommendations on the level of Goldwater-Nichols, Nunn-Cohen, or Lieberman-Collins for (potential) necessary changes in funding, structures, and authorities for irregular warfare. These previous efforts demonstrated the critical role of the U.S. Congress in forcing change in the wake of strategic failure by the United States, compelling the

---

[5] The importance of congressional involvement is emphasized by the failure of internal initiatives, such as the Joint Low-Intensity Conflict Project (which was directed by the Army chief of staff), to drive any significant reform from within. The final report from this project contains many elements of what I believe should be part of the formal congressional inquiry (TRADOC, *Joint Low-Intensity Conflict Project: Final Report*, Fort Monroe, Va., 1987).

U.S. national security community to form SOCOM and the National Counterterrorism Center.

However, there is a major difference between irregular warfare and these previous congressionally mandated reforms, in that we do not have a systematic understanding of why we are failing in these conflicts. Mandating the creation of another command structure dedicated to this form of warfare, mimicking the creation of SOCOM by Nunn-Cohen, is one possibility.

I suspect that a thorough review of our strategic failures in these contests may result in broader recommended reforms to our national security enterprise. In particular, I have come to believe that there is a need for the United States to formalize and develop what might be best called *irregular statecraft*. Irregular statecraft is a form of competition in which state and nonstate actors employ all means, short of war, to support friends and allies and erode the influence, legitimacy, and authority of adversaries and is the modern equivalent of what George Kennan described, in 1948, as *political warfare*.[6]

Irregular statecraft encompasses offensive and defensive capabilities, both covert and overt, currently dispersed across the U.S. government, including the irregular warfare capability of the U.S. military and complementary capabilities among a multitude of diplomatic, intelligence, homeland security, and other organizations. Existing U.S. expertise in the contemporary use of irregular statecraft is disparate and dormant and would require congressional action to aggregate experts from outside government and representatives from across relevant U.S. agencies to synchronize and mature America's expertise in this form of global competition.

---

[6] George F. Kennan, *Policy Planning Staff Memorandum 269*, Washington, D.C.: U.S. Department of State, May 4, 1948. Others have come to a similar conclusion about this need (e.g., Linda Robinson, Todd C. Helmus, Raphael S. Cohen, Alireza Nader, Andrew Radin, Madeline Magnuson, and Katya Migacheva, *Modern Political Warfare: Current Practices and Possible Responses*, Santa Monica, Calif.: RAND Corporation, RR-1772-A, 2018). However, I have learned that the term *political warfare* is deeply problematic to many outside the Department of Defense and does not fully capture the type of capability that I believe that the nation needs to build.

Recommendation to the President: Reorganize the Executive Branch Around the Security Challenges of the 21st Century

The U.S. national security enterprise within the Executive Branch is a legacy operating system designed to fight the Great Power competitions of the 20th century. We remain peerless in our conventional military capabilities yet find ourselves vulnerable to the irregular strategies of terrorists, insurgents, and our Great Power adversaries. Reform from within is unlikely, as the dominant view within the Department of Defense is that success in Great Power competition requires overmatch in conventional, cyber, and nuclear. This is true, but insufficient, as it leaves the United States vulnerable to irregular strategies, as we have seen in Crimea, Yemen, and elsewhere.[7]

Maturing the American way of irregular war will very likely require reform of the Executive Branch. Although the recommended congressional study (if implemented) might provide somewhat different recommendations, I believe that the President of the United States and staff ought to examine options for how the United States could better organize itself for success in these conflicts.

There are three options worthy of consideration, described immediately below, each of which embraces the inherently interagency and joint nature of American irregular warfare that I saw during my career. Although I do believe that U.S. Army Special Forces will continue to play a prominent role in irregular warfare, the American way of irregular war must incorporate military, foreign service, and intelligence professionals in prominent (if not leadership) roles, and the U.S. special operation's community (as is currently the case) might not necessarily be best positioned to own this mission. Each of these options recognizes that research and analysis will be necessary to develop concepts for America's use of irregular warfare and that maintaining a persistent

---

[7] In retrospect, this is not so dissimilar for Peter Paret's characterization of the challenge that Napoleon France faced in using its *levée en masse* to fight small wars that I read about at West Point at the very beginning of my career (Peter Paret, "Napoleon and the Revolution in War," in Peter Paret, Gordon A. Craig, and Felix Gilbert, eds., *Makers of Modern Strategy from Machiavelli to the Nuclear Age*, Princeton, N.J.: Princeton University Press, 1986).

but low-profile global network of irregular warfare professionals will be critical to providing situational awareness; delivering access, influence, and indigenous options; and enhancing America's use of influence operations.[8]

The first option is to create a contemporary Office of Strategic Services by establishing a cabinet-level organization with primary responsibility for paramilitary, influence, and special warfare operations. During World War II, the Office of Strategic Services provided a strategic-level irregular warfare capability that combined special operations and intelligence expertise. This hybrid civilian-military agency would be led by a combination of political appointees, military personnel, intelligence professionals, and foreign service professionals. It would take responsibility for all irregular warfare activities that are currently in the domain of the Department of Defense, Department of State, and CIA. This would allow the CIA to focus on strategic intelligence, the Department of Defense to focus on deterring and (if necessary) fighting traditional wars, and the Department of State to focus on diplomacy.

A second option would be to create a separate service within the Department of Defense that is missioned to own irregular warfare. This new service would focus on understanding and successfully operating in the human domain and develop concepts, doctrine, personnel, and structures for success against resistance and insurgency. It would take primary responsibility for foreign internal defense (supporting friendly governments in defending against resistance movements) and unconventional warfare (supporting resistance movements against our adversaries) and would field formations and headquarters that conduct the range of interactions with foreign militaries, militias, and surrogate

---

[8]   Each of these three options creates a new superstructure for irregular warfare that might have the unintended consequence of reducing flexibility for our tactical-level irregular warriors because of increased oversight, an effect similar to the loss of "specialness" for U.S. special operations forces following the creation of SOCOM (Jessica Turnley, *Retaining a Precarious Value as Special Operations Go Mainstream*, MacDill Air Force Base, Fla.: Joint Special Operations University Press, 2008). The challenge here is somewhat different, as I firmly believe that conventional forces and the interagency have important roles to play in the American way of irregular war. But allowing flexibility while legitimizing what these irregular warriors are doing will be an ongoing challenge.

groups. This would allow the Army and the Marine Corps to focus on dominating in the physical component of the land domain, their intended purpose and for which they are best designed, and provide geographic combatant commands with headquarters, warfighters, and assets necessary to execute theater-level irregular warfare campaigns. A variant of this second option would be to create a new service under the Department of the Army that is focused on irregular warfare, building off the model that is currently employed by the Navy and Marine Corps.

A third option, which, like the second option, has been proposed many times before, would be to divide SOCOM into two functional commands. The first would focus on the national-priority missions of counterterrorism and countering weapons of mass destruction, most closely aligning with SOCOM's current focus. The second functional command would focus on building global networks among friends and allies to better enable them to fight their own battles against insurgencies, overthrow their oppressors, or even deter authoritarian Great Powers that survive by oppressing their peoples. Each resulting four-star functional combatant command would be provided its own major funding program line authorized and appropriated by Congress.

In this arrangement, Joint Special Operations Command would merge with the counterterrorism-focused functional command, and the six theater special operation commands (e.g., SOCCENT, SOCSOUTH) would become standing two- or three-star component commands for the new irregular warfare–focused oriented functional command. Operational control of the theater special operation commands would return to the geographic combatant commands,[9] and the four-star irregular warfare command would be responsible for manning, training, and equipping the forces that fill these irregular warfare–focused regional headquarters.

This third approach would be less disruptive than the first two, in that it would allow the services to remain force providers and would capitalize on a positive trend in at least the Army and the Marine

---

[9] The theater special operation commands were realigned under SOCOM in 2013 (Michael D. Tisdel, Ken D. Teske, and William C. Fleser, *Theater Special Operations Commands Realignment*, MacDill Air Force Base, Fla.: U.S. Special Operations Command, 2014).

Corps to more fully embrace their irregular warfare communities. As an example, USASOC would continue to provide forces for both new functional combatant commands and would draw funds from each, as well as the Army to man, train, and equip its forces. The other current SOCOM service components could provide forces for each as well, depending on the mission.

Recommendation to Concerned Citizens: Establish an Institution Outside Government Dedicated to Understanding American Irregular Warfare

Arguably, our victory in the Cold War resulted as much from America's ability to harness its intellectual capacity as it did from its industrial might. Academic institutions and think tanks became vital in expanding U.S. thinking on defense and in providing experts to serve in government. These patriots then provided the intellectual underpinning for policies and strategies that kept communism at bay and eventually won the Cold War.

The United States has not had—again, in my view and based on my own experience—the same quality of expertise for irregular warfare. There are pockets of excellence within the U.S. government that study the phenomenon of irregular warfare, such as SOCOM's Joint Special Operations University and the Naval Postgraduate School's Department of Defense Analysis and Special Operations/Irregular Warfare program. And there are a handful of academics at universities and think tanks dedicated to the study of aspects of irregular warfare. However, this existing expertise is limited in that few professionals can dedicate themselves to the study of irregular warfare in any meaningful way. Further, as consequence of how these organizations are funded, few are in a position to provide an independent and critical perspective on America's performance in this form of warfare.[10]

---

[10] In addition, the work of these few academics often falls on hard ground, and results in little change, as there is no organization within the U.S. government that necessarily cares about or drives research. That said, even if my first two recommendations are not adopted, which would create the irregular warfare–focused organizational structures, I do believe that this institution could create necessary momentum for change.

I believe that an independently funded center, or a public-private center supported by both Congress and private citizens, at a university or think tank dedicated to the study of American irregular warfare would provide our country three necessary capabilities. The first is a continuous and independent critique of U.S. capabilities, policies, and strategies in irregular warfare to determine how the United States is performing in its many irregular engagements and might do better in this type of war. Second, it would provide a stable of professionals who are expert in the contemporary use of irregular warfare, both how our adversaries apply this form of warfare and how the United States can deploy irregular warfare defensively and offensively to contest these adversaries, whether state or nonstate. The third would be to capture and analyze irregular warfare experiences, providing a publicly available record of our successes and failures and thus serving as a bridge between irregular warfare practitioners and the American people.[11]

This center, which might be established in concert with the recommended congressional review and combine a broad range of different expertise, could be focused on irregular warfare or perhaps on the broader concept of irregular statecraft.[12] It would be directed by an internationally recognized national security professional who would be essential for both guiding the research and engaging with Congress, the Department of Defense, the Department of State, the CIA, and other diplomatic, intelligence, and homeland security organizations. The staff would include a modest number of academics representing the diverse skill sets necessary for irregular warfare (or irregular statecraft) at the operational and strategic levels, including anthropology, economics, psychology, and sociology, but also practitioners from across relevant U.S. agencies. This center would likely prove a worthy investment of both private and public funds, as it reduces the risk of costly interventions and makes the United States more likely to gain an enduring advantage in global competition.

---

[11] We thank James Kiras for this excellent suggestion.

[12] This could be patterned after the European Centre of Excellence for Countering Hybrid Threats, which is designed to study and design defenses against modern hybrid threats.

CHAPTER FOURTEEN

Conclusion

Competition with other nations is the primary national security concern of the United States.[1] This includes competition with the revisionist powers of China and Russia,[2] the "central challenge to U.S. prosperity and security,"[3] and rogue regimes (e.g., Iran, North Korea) that threaten to destabilize regions through "their pursuit of nuclear weapons or sponsorship of terrorism."[4] Deterring the aggression of these adversaries will certainly require that the United States expand the lethality of its conventional and nuclear capabilities to sustain its military supremacy.[5]

However, actual competition and conflict with these nation-states will most likely be irregular.[6] This reflects a deliberate calculus by our

---

[1] U.S. Department of Defense, *Summary of the 2018 National Defense Strategy of the United States of America: Sharpening the American Military's Competitive Edge*, Washington, D.C., January 19, 2018, p. 1; Daniel R. Coats, *Worldwide Threat Assessment of the U.S. Intelligence Community*, statement for the Senate Select Committee on Intelligence, Washington, D.C.: Office of the Director of National Intelligence, January 29, 2019, p. 4.

[2] For a discussion of this Great Power competition, see Graeme P. Herd, *Great Powers and Strategic Stability in the Twenty-First Century: Competing Visions of World Order*, Abingdon, UK: Routledge, 2010.

[3] U.S. Department of Defense, 2018, p. 2.

[4] U.S. Department of Defense, 2018, p. 2.

[5] This is described as "Build a More Lethal Force" in the National Defense Strategy (U.S. Department of Defense, 2018).

[6] These nations will seek to "exert influence over the politics and economies of states in all regions of the world and especially in their respective backyards" (Coats, 2019, p. 4).

adversaries, who understand that "conventional or nuclear war with the United States would be risky and prohibitively costly,"[7] but that the United States is vulnerable to irregular approaches. Rather than risk conflicts with casualty numbers that are "virtually unthinkable," these adversaries will continue to deploy a blend of information, legal, and proxy warfare in challenging the United States and its allies.[8] These irregular approaches allow America's adversaries to turn the "democratic norms and institutions" of the United States and its allies against them, allowing for contestation without direct military action.[9]

Maturing the American way of irregular war is critical for the United States to remain competitive against these threats. We need to be as agile as our adversaries, who are focused on competition and conflict short of traditional war and have already demonstrated its potency against the United States and its allies. And we must have the capability to be both reactive and proactive, allowing us to simultaneously counter our adversaries' irregular threats and go on the offensive. This offensive capability will give the United States an added ability to deter our enemies by expanding the competitive space of the U.S. military.[10] The Achilles' heel of our authoritarian adversaries is their inherent fear of their own people; the United States must be ready to capitalize on this fear.

However, I sense in my interaction with those still in government—uniform and civilian—that there is a growing desire to get back to business as usual. They want to move beyond the professional embarrassment represented by the inconclusive results in Afghanistan, the muddle that is the conflicts in Iraq and Syria, the emergence of the

---

[7]   Seth Jones, "The Future of Irregular Warfare Is Irregular," *National Interest*, August 26, 2018.

[8]   Jones, 2018; Valery Gerasimov, "The Value of Science Is in the Foresight: New Challenges Demand Rethinking the Forms and Methods of Carrying Out Combat Operations," *Military Review*, January–February 2016.

[9]   Mark Galeotti, "I'm Sorry for Creating the 'Gerasimov Doctrine,'" *Foreign Policy*, March 5, 2018.

[10]   U.S. Department of Defense, 2018, p. 4; Terrence J. O'Shaughnessy, Matthew D. Strohmeyer, and Christopher D. Forrest, "Strategic Shaping: Expanding the Competitive Space," *Joint Forces Quarterly*, Vol. 90, No. 3, July 2018.

Islamic State, and the proliferation and metastasis of al Qaeda and its brethren. With the growing threats from rival powers demanding that U.S. military capabilities be rededicated to their traditional missions, it is likely that the United States will return to conventional approaches to conflict without outside pressure to do otherwise.

There is a real risk that we repeat the mistakes that we made in the wake of Vietnam—specifically, that we again fail to take seriously the lessons of past conflicts to adapt to a changing threat landscape. There is no question that readiness in the Army and Marine Corps has suffered in recent years because of the focus on counterinsurgency and partnered warfare. And the services will find it increasingly difficult to remain dominant in traditional and nuclear war to ensure deterrence, while maintaining the proficiency and resolve to win in these irregular contests. For many, the return to traditional war and the pivot from these contests against adversaries wielding nonconventional means are much welcomed. It is their comfort zone, after all, and where the United States has dominated for a hundred years.

The U.S. Congress, the President of the United States, and the American people should not and cannot assume that the Department of Defense will either sustain the irregular warfare capabilities it has developed over these past decades or mature the additional capabilities it needs to be competitive against these irregular adversaries. In the past, it has taken the initiative and persistence of our civilian leaders— in Congress and in the White House—to force the Department of Defense to make the reforms necessary to remain lethal against a modern adversary. Further, as I hope that I have convinced you in this narrative through my own experience, the American way of irregular war relies on many capabilities that lie outside the Department of Defense. The American way of irregular war is an inherently whole-of-government enterprise, requiring "seamless integration of multiple elements of national power—diplomacy, information, economics, finance, intelligence, law enforcement, and military."[11]

An American way of irregular war will reflect who we are as a people, our diversity, our moral code, and our undying belief in free-

---

[11] U.S. Department of Defense, 2018, p. 4.

dom and liberty. It must be both defensive and offensive. Developing it will take time, will require support from the American people through their Congress, and is guaranteed to disrupt the status quo and draw criticism. It will take leadership, dedication, and courage. It is my hope that this study encourages, informs, and animates those with responsibility to protect the nation to act. Our adversaries have moved to dominate in the space below the threshold of war. It will be a strategy built around an American way of irregular war that defeats them.

References

Abbott, Michael H., "The Army and the Drug War: Politics or National Security?" *Parameters*, December 1988, pp. 95–112.

Adams, Thomas K., *U.S. Special Operations Forces in Action: The Challenge of Unconventional Warfare*, Portland, Ore.: Frank Cass Publishers, 1998.

Air Force Association, *Strategy, Requirements, and Forces: The Rising Imperative of Air and Space Power*, Arlington, Va., 2003. As of January 6, 2019:
http://www.dtic.mil/dtic/tr/fulltext/u2/a466440.pdf

Americas Watch Committee and American Civil Liberties Union, *Report on Human Rights in El Salvador*, New York: Vintage, 1982.

Army Doctrine Publication 3-05, *Special Operations*, Washington, D.C.: Headquarters, Department of the Army, August 2012.

Army Techniques Publication 3-07.5, *Stability Techniques*, Washington, D.C.: Headquarters, U.S. Department of the Army, August 2012. As of January 30, 2020:
https://armypubs.army.mil/epubs/DR_pubs/DR_a/pdf/web/atp3_07x5.pdf

Averett, Christian M., Louis A. Cervantes, and Patrick M. O'Hara, "An Analysis of Special Operations Command—South's Distributive Command and Control Concept," thesis, Monterey, Calif.: Naval Postgraduate School, 2007.

Bacevich, Andrew J., James D. Hallums, Richard H. White, and Thomas F. Young, *American Military Policy in Small Wars: The Case of El Salvador*, Washington, D.C.: Pergamon-Brassey's, 1988.

Ball, Tim, "Bringing the Alliance Back to SOF: The Role of NATO Special Operations Headquarters in Countering Russian Hybrid Warfare," in *2016 Special Operations Essays*, MacDill Air Force Base, Fla.: Joint Special Operations University Press, 2016, pp. 13–23.

Barry, John, "America's Secret Libya War: U.S. Spent $1 Billion on Covert Ops Helping NATO," *Daily Beast*, July 13, 2017.

Bartkus, Viva, "'Untapped Resources' for Building Security from the Ground Up," *Joint Forces Quarterly*, Vol. 93, No. 2, May 2019.

Basquez, Andrew, "SF Returns to Its Roots with the 4th Battalion Redesign," *Special Warfare*, October–December 2013.

Beckwith, Charles, and Donald Knox, *Delta Force*, New York: Harcourt Brace Javanovich Publishers, 1983.

Best, Richard A., Jr., and Andrew Feickert, *Special Operations Forces (SOF) and CIA Paramilitary Operations: Issues for Congress*, Washington, D.C.: Congressional Research Service, 2006.

Blais, Curtis, *Governance Innovation for Security and Development: Recommendations for US Army Civil Affairs 38G Civil Sector Officers*, Monterey, Calif.: Naval Postgraduate School, 2014.

Bolduc, Donald C., "The Future of Afghanistan," *Special Warfare*, October–December 2011.

Boot, Max, "The New American Way of War," *Foreign Affairs*, July–August, 2003.

Boykin, William G., *Special Operations and Low-Intensity Conflict Legislation: Why Was It Passed and Have the Voids Been Filled?* Carlisle, Pa.: U.S. Army War College, 1991.

Boyne, Walter J., "The Fall of Saigon," *Air Force Magazine*, April 2000.

Brennan, Richard R., Jr., Charles P. Ries, Larry Hanauer, Ben Connable, Terrence K. Kelly, Michael J. McNerney, Stephanie Young, Jason Campbell, and K. Scott McMahon, *Ending the U.S. War in Iraq: The Final Transition, Operational Maneuver, and Disestablishment of United States Forces–Iraq*, Santa Monica, Calif.: RAND Corporation, RR-232-USFI, 2013. As of April 28, 2020: https://www.rand.org/pubs/research_reports/RR232.html

Brigg, Morgan, and Roland Bleiker, "Autoethnographic International Relations: Exploring the Self as a Source of Knowledge," *Review of International Studies*, Vol. 36, No. 3, 2010, pp. 779–798.

Brinkley, Joel, "The Talk of Trinidad; Bolivian Town Resents Drug Glare," *New York Times*, July 25, 1986.

Briscoe, Charles H., and Daniel J. Kulich, "Operación Jaque: The Ultimate Deception," *Veritas*, Vol. 14, No. 3, 2018.

Brooks, Drew, "Lt. Gen. Cleveland: Changing World Requires New Mission for Special Operations," *Fayetteville Observer*, March 18, 2015. As of November 12, 2018: https://www.fayobserver.com/article/20150318/News/303189724

Brown, Bryan D., "U.S. Special Operations Command Meeting the Challenges of the 21st Century," *Joint Forces Quarterly*, Vol. 40, No. 1, 2006, pp. 38–43.

Brown, Timothy D., *Unconventional Warfare as a Strategic Force Multiplier: Task Force Viking in Northern Iraq, 2003*, MacDill Air Force Base, Fla.: Joint Special Operations University Press, September 2017.

Bush, George H. W., "Remarks at Maxwell Air Force Base War College in Montgomery, Alabama," April 13, 1991, transcript from Gerhard Peters and John T. Woolley, American Presidency Project. As of January 6, 2019: https://www.presidency.ucsb.edu/node/265992

———, "Address to the United Nations General Assembly in New York City," September 21, 1992, transcript from Gerhard Peters and John T. Woolley, American Presidency Project. As of January 6, 2019: https://www.presidency.ucsb.edu/documents/ address-the-united-nations-general-assembly-new-york-city

Buswell, Philip A., "Keeping Special Forces Special: Regional Proficiency in Special Forces," thesis, Monterey, Calif.: Naval Postgraduate School, 2011.

Cale, Paul, "The United States Military Advisory Group in El Salvador, 1979–1992," *Small Wars Journal*, 1996.

Cannady, Bryan, "Irregular Warfare: Special Operations Joint Professional Military Education Transformation," thesis, Fort Leavenworth, Kan.: U.S. Army Command and General Staff College, 2008.

Cavanaugh, Matt, "Strategic Landpower Is Dead: Long Live Strategic Landpower," Modern War Institute, August 14, 2016. As of April 30, 2020: https://mwi.usma.edu/strategic-landpower-dead-long-live-strategic-landpower/

Central Intelligence Agency, "Surprise, Kill, Vanish: The Legend of the Jedburghs," December 17, 2015. As of January 31, 2020: https://www.cia.gov/news-information/featured-story-archive/2015-featured-story-archive/legend-of-the-jedburghs.html

Centro Documental Historico Militar, "Batallon Bracamonte," webpage, undated. As of February 26, 2017: http://miriammixco.com/?q=node/609

Childress, Michael, *The Effectiveness of U.S. Training Efforts in Internal Defense and Development: The Cases of El Salvador and Honduras*, Santa Monica, Calif.: RAND Corporation, MR-250-USDP, 1995. As of April 30, 2020: https://www.rand.org/pubs/monograph_reports/MR250.html

CIA—*See* Central Intelligence Agency.

Clary, Mike, "Key Witness Against Noriega Sentenced to Time Served," *Los Angeles Times*, July 10, 1992.

Cleveland, Charles T., *Command and Control of the Joint Commission Observer Program—U.S. Army Special Forces in Bosnia*, Carlisle, Pa.: U.S. Army War College, 2001.

Cleveland, Charles T., Ryan Crocker, Daniel Egel, Andrew M. Liepman, and David Maxwell, *An American Way of Political Warfare: A Proposal*, Santa Monica, Calif.: RAND Corporation, PE-304, 2018. As of April 30, 2020: https://www.rand.org/pubs/perspectives/PE304.html

Cleveland, Charles T., James B. Linder, and Ronald Dempsey, "Special Operations Doctrine: Is It Needed?" *Prism*, Vol. 6, No. 3, 2016.

CNN, "U.S. Reinforcements Arrive in Kirkuk," April 10, 2003.

Coats, Daniel R., *Worldwide Threat Assessment of the U.S. Intelligence Community*, statement for the Senate Select Committee on Intelligence, Washington, D.C.: Office of the Director of National Intelligence, January 29, 2019.

Cole, Ronald H., *Operation Just Cause: Panama*, Washington, D.C.: Joint History Office, 1995.

Congressional Budget Office, *Reduction in the Army Officer Corps*, Washington, D.C., April 1992. As of February 13, 2020:
https://apps.dtic.mil/dtic/tr/fulltext/u2/a529435.pdf

Congressional Research Service, "U.S. War Costs, Casualties, and Personnel Levels Since 9/11," Washington, D.C., April 18, 2019. As of February 15, 2020:
https://fas.org/sgp/crs/natsec/IF11182.pdf

Connable, Ben, *Redesigning Strategy for Irregular War: Improving Strategic Design for Planners and Policymakers to Help Defeat Groups Like the Islamic State*, Santa Monica, Calif.: RAND Corporation, WR-1172-OSD, 2017. As of February 14, 2020:
https://www.rand.org/pubs/working_papers/WR1172.html

Cordesman, Anthony, "The Real Revolution in Military Affairs," Center for Strategic and International Studies, August 5, 2014. As of September 3, 2018:
https://www.csis.org/analysis/real-revolution-military-affairs

Cornelius, Robert L., *An Evaluation of the Human Domain Concept: Organizing the Knowledge, Influence, and Activity in Population-Centric Warfare*, Fort Leavenworth, Kan.: School of Advanced Military Studies, U.S. Army Command and General Staff College, 2015.

Craig, Richard B., "Operation Intercept: The International Politics of Pressure," *Review of Politics*, Vol. 42, No. 4, 1980, pp. 556–580.

Crandall, Russel, *America's Dirty Wars: Irregular Warfare from 1776 to the War on Terror*, New York: Cambridge University Press, 2014.

Dale, Catherine, *Operation Iraqi Freedom: Strategies, Approaches, Results, and Issues for Congress*, Washington, D.C.: Congressional Research Service, March 28, 2008.

Daugherty, William J., *Executive Secrets: Covert Action and the Presidency*, Lexington: University Press of Kentucky, 2004.

Davies, Michael C., and Frank Hoffman, "Joint Force 2020 and the Human Domain: Time for a New Conceptual Framework?" *Small Wars Journal*, June 2013. As of December 7, 2018:
http://smallwarsjournal.com/jrnl/art/
joint-force-2020-and-the-human-domain-time-for-a-new-conceptual-framework

Defense Intelligence Agency, *Russia Military Power*, Washington, D.C., 2017. As of January 30, 2020:
https://www.dia.mil/portals/27/documents/news/military%20power%20
publications/russia%20military%20power%20report%202017.pdf

De Luce, Dan, "Meet the Venture Capitalist Who Launched a Kickstarter for War," *Foreign Policy*, May 31, 2017. As of February 22, 2019:
https://foreignpolicy.com/2017/05/31/
meet-the-venture-capitalist-who-launched-a-kickstarter-for-war/

Department of Defense Directive 5105.75, *Department of Defense Operations at U.S. Embassies*, Washington, D.C.: U.S. Department of Defense, December 21, 2007. As of February 13, 2020:
https://www.hsdl.org/?view&did=482023

DeYoung, Karen, "U.S. Withholding Military Aid to Pakistan," *Washington Post*, July 10, 2011. As of November 1, 2018:
https://www.washingtonpost.com/world/asia-pacific/us-withholding-military-aid-to-pakistan/2011/07/10/gIQAZdJH7H_story.html

Dreyer, June Teufel, "People's Liberation Army Lessons from Foreign Conflicts: The Air War in Kosovo," in Andrew Scobell, David Lai, and Roy Kamphausen, eds., *Chinese Lessons from Other Peoples' Wars*, Carlisle, Pa.: Strategic Studies Institute, U.S. Army War College, 2011.

Echevarria, Antulio J., II, "What Is Wrong with the American Way of War?" *Prism*, Vol. 3, No. 4, 2012, pp. 109–115. As of January 25, 2020:
https://cco.ndu.edu/Portals/96/Documents/prism/prism_3-4/prism108-115_echevarria.pdf

Eck, Kristine, and Thérése Pettersson, "Organized Violence, 1989–2017 and the Data Generation Process," *Journal of Peace Research*, Vol. 55, No. 4, 2018.

European Union, "EUFOR Operation ALTHEA European Union Military Operation Bosnia and Herzegovina," fact sheet, February 2020.

Evans, Ernest, "El Salvador's Lessons for Future U.S. Interventions," *World Affairs*, Vol. 160, No. 1, 1997, pp. 43–48.

"Excerpts from Haig's Briefing About El Salvador," *New York Times*, February 21, 1981.

Farthing, Linda, and Benjamin Kohl, "Social Control: Bolivia's New Approach to Coca Reduction," *Latin American Perspectives*, Vol. 37, No. 4, 2010.

Feickert, Andrew, *U.S. Special Operations Forces (SOF): Background and Issues for Congress*, Washington, D.C.: Congressional Research Service, July 16, 2010.

Field Manual 3-18, *Special Forces Operations*, Washington, D.C.: Headquarters, U.S. Department of the Army, May 2014.

Field Manual 3-24 and Marine Corps Warfighting Publication 3-33.5, *Counterinsurgency*, Washington, D.C.: U.S. Department of the Army, December 16, 2006.

Field Manual 3-24 and Marine Corps Warfighting Publication 3-33.5, *Insurgencies and Countering Insurgencies*, Washington, D.C.: U.S. Department of the Army, May 2014.

Field Manual 31-21, *Special Forces Operations*, Washington, D.C.: Headquarters, U.S. Department of the Army, June 1965.

Field Manual 100-5, *Operations*, Washington, D.C.: Headquarters, Department of the Army, June 1993.

Findlay, Michael L., *Special Forces Integration with Multinational Division–North in Bosnia-Herzegovina*, Fort Leavenworth, Kan.: School of Advanced Military Studies, U.S. Army Command and General Staff College, 1998.

Fishel, John T., *The Fog of Peace: Planning and Executing the Restoration of Panama*, Carlisle, Pa.: Strategic Studies Institute, U.S. Army War College, April 15, 1992. As of January 30, 2020:
https://apps.dtic.mil/dtic/tr/fulltext/u2/a251124.pdf

Freeman, Mark P., "Autobiography," in Lisa M. Given, ed., *The SAGE Encyclopedia of Qualitative Research Methods*, Vol. 2, Thousand Oaks, Calif.: SAGE Publications, 2008, pp. 45–48.

Fulwood, Sam, III, "Combat in Panama: Dignity Battalion Still Lurks in City Shadows," *Los Angeles Times*, December 22, 1989. As of February 10, 2020:
https://www.latimes.com/archives/la-xpm-1989-12-22-mn-705-story.html

Galeotti, Mark, "I'm Sorry for Creating the 'Gerasimov Doctrine,'" *Foreign Policy*, March 5, 2018. As of October 8, 2018:
https://foreignpolicy.com/2018/03/05/
im-sorry-for-creating-the-gerasimov-doctrine/

Galvin, John R., "Uncomfortable Wars: Toward a New Paradigm," *Parameters*, Vol. 16, No. 4, 1986. As of February 6, 2018:
http://www.dtic.mil/dtic/tr/fulltext/u2/a521482.pdf

Ganser, Daniele, "Terrorism in Western Europe: An Approach to NATO's Secret Stay-Behind Armies," *Whitehead Journal of Diplomacy and International Relations*, Winter–Spring 2005, pp. 69–95.

Gerasimov, Valery, "The Value of Science Is in the Foresight: New Challenges Demand Rethinking the Forms and Methods of Carrying Out Combat Operations," *Military Review*, January–February 2016.

Gleditsch, Nils Petter, Peter Wallensteen, Mikael Eriksson, Margareta Sollenberg, and Håvard Strand, "Armed Conflict 1946–2001: A New Dataset," *Journal of Peace Research*, Vol. 39, No. 5, 2002, pp. 615–637.

Gleijese, Piero, *Conflicting Missions: Havana, Washington, and Africa, 1959–1976*, Chapel Hill, University of North Carolina Press, 2002.

Gleiman, Jan Kenneth, *Operational Art and the Clash of Organizational Cultures: Postmortem on Special Operations as a Seventh Warfighting Function*, Fort Leavenworth, Kan.: School of Advanced Military Studies, U.S. Army Command and General Staff College, 2011. As of February 10, 2020:
https://pdfs.semanticscholar.org/27ce/26fae6ac7abcdbdb258eb8d83cbf0e3bd184.pdf

Gordon, Michael, "General Says Salvador Can't Defeat Guerrillas," *New York Times*, February 9, 1990.

Gray, Colin S., "Irregular Warfare: One Nature, Many Characters," *Strategic Studies Quarterly*, Vol. 1, No. 2, 2007, pp. 35–57.

Gray, John, "Our Newest Protectorate," *The Guardian*, April 26, 1999. As of January 30, 2020:
https://www.theguardian.com/world/1999/apr/27/balkans9

Greentree, Todd, "Bureaucracy Does Its Thing: US Performance and the Institutional Dimension of Strategy in Afghanistan," *Journal of Strategic Studies*, Vol. 36, No. 3, 2013, pp. 325–356.

Grenier, John, *The First Way of War: American War Making on the Frontier, 1607–1814*, Cambridge: Cambridge University Press, 2008.

Harclerode, Peter, *Fighting Dirty: The Inside Story of Covert Operations from Ho Chi Minh to Osama Bin Laden*, London: Cassell and Company, 2001.

Heisler, Anthony F., "By, with, and Through: The Theory and Practice of Special Operations Capacity-Building," thesis, Monterey, Calif.: Naval Postgraduate School, 2014.

Hennelly, Michael J., "US Policy in El Salvador: Creating Beauty or the Beast?" *Parameters*, Spring 1993, pp. 59–69.

Herd, Graeme P., *Great Powers and Strategic Stability in the Twenty-First Century: Competing Visions of World Order*, Abingdon, UK: Routledge, 2010.

Hilton, Corson L., *United States Army Special Forces: From a Decade of Development to a Sustained Future*, Carlisle, Pa.: U.S. Army War College, 1991.

Hutchens, Michael E., William D. Dries, Jason C. Perdew, Vincent D. Bryant, and Kerry E. Moores, "Joint Concept for Access and Maneuver in the Global Commons: A New Joint Operational Concept," *Joint Forces Quarterly*, Vol. 84, No. 1, 2017.

Hutchinson, David S., *The 3d Battalion 27th Infantry in Operation Just Cause*, Carlisle, Pa.: U.S. Army War College, 1992. As of June 8, 2020:
https://apps.dtic.mil/dtic/tr/fulltext/u2/a296093.pdf

Inciardi, James A., *Handbook of Drug Control in the United States*, New York: Greenwood Press, 1990.

International Crisis Group, *Brcko: What Bosnia Could Be*, Brussels, February 10, 1998.

Isikoff, Michael, "DEA in Bolivia 'Guerilla Warfare,'" *Washington Post*, January 16, 1989.

Jackson, Jose F., "'A Just Cause,'" *Citizen Airman*, March 1990, pp. 30–31.

Jannarone, August G., and Ray E. Stratton, "Toward an Integrated United States Strategy for Counternarcotics and Counterinsurgency," *DISAM Journal*, Winter 1990–1991.

Jensen, Jack J., "Special Operations Command (Forward)—Lebanon: SOF Campaigning 'Left of the Line,'" *Special Warfare*, April–June 2012.

Johnson, William M., *U.S. Army Special Forces in Desert Shield/Desert Storm: How Significant an Impact*, Fort Leavenworth, Kan.: U.S. Army Command and General Staff College, 1996.

Joint Publication 3-05, *Special Operations*, Washington, D.C.: Joint Chiefs of Staff, 2003.

Jones, D., *Ending the Debate: Unconventional Warfare, Foreign Internal Defense, and Why Words Matter*, Fort Leavenworth, Kan.: U.S. Army Command and General Staff College, 2006.

Jones, Seth, "The Future of Irregular Warfare Is Irregular," *National Interest*, August 26, 2018. As of October 5, 2018:
https://nationalinterest.org/feature/future-warfare-irregular-29672

Kavanagh, Jennifer, Bryan Frederick, Alexandra Stark, Nathan Chandler, Meagan L. Smith, Matthew Povlock, Lynn E. Davis, and Edward Geist, *Characteristics of Successful U.S. Military Interventions*, Santa Monica, Calif.: RAND Corporation, RR-3062-A, 2019. As of January 22, 2020:
https://www.rand.org/pubs/research_reports/RR3062.html

Kendrick, Scott, "Review of *Force Without War: U. S. Armed Forces as a Political Instrument and Implications* for Future Doctrine and Education," white paper, Strategic Landpower Task Force, July 31, 2017. As of January 22, 2020:
http://arcic-sem.azurewebsites.us/App_Documents/SLTF/Review-of-Force-without-War.pdf

Kennan, George F., *Policy Planning Staff Memorandum 269*, Washington, D.C.: U.S. Department of State, May 4, 1948.

Kennedy, John F., *Special Warfare*, Washington, D.C.: Office of the Chief of Information, Department of the Army, 1962.

Kent, Glenn A., with David Ochmanek, Michael Spirtas, and Bruce R. Pirnie, *Thinking About America's Defense: An Analytical Memoir*, Santa Monica, Calif.: RAND Corporation, OP-223-AF, 2008. As of May 7, 2020: https://www.rand.org/pubs/occasional_papers/OP223.html

Kiras, James, *Special Operations and Strategy: From World War II to the War on Terrorism*, New York: Routledge, 2006.

―――, "A Theory of Special Operations: 'These Ideas Are Dangerous,'" *Special Operations Journal*, Vol. 1, No. 2, 2015, pp. 75–88.

Komer, Robert W., *Bureaucracy Does Its Thing: Institutional Constraints on U.S.-GVN Performance in Vietnam*, Santa Monica, Calif.: RAND Corporation, R-967-ARPA, 1972. As of May 7, 2020: https://www.rand.org/pubs/reports/R967.html

Lacina, Bethany, and Nils Petter Gleditsch, "Monitoring Trends in Global Combat: A New Dataset of Battle Deaths," *European Journal of Population*, Vol. 21, Nos. 2–3, 2005, pp. 145–166.

Lane, William R., *Resourcing for Special Operations Forces (SOF) Should Responsibilities Be Passed from USSOCOM Back to the Services*, Carlisle, Pa.: U.S. Army War College, 2006.

Latal, Srecko, "Bosnia's Dayton Deal Is Dead—So What Now?" *Balkan Insight*, December 2, 2019. As of February 14, 2020: https://balkaninsight.com/2019/12/02/bosnias-dayton-deal-is-dead-so-what-now/

Lehman, Joshua, *Leading in the Gray Zone: Command and Control of Special Operations in Phases 0-1*, Naval Station Newport, R.I.: Naval War College, 2016.

Livingston, Ian S., and Michal O'Hanlon, *Afghanistan Index*, Washington, D.C.: Brookings Institution, May 25, 2017. As of January 17, 2020: https://www.brookings.edu/wp-content/uploads/2016/07/21csi_20170525_afghanistan_index.pdf

Loredo, Elvira N., John E. Peters, Karlyn D. Stanley, Matthew E. Boyer, William Welser IV, and Thomas S. Szayna, *Authorities and Options for Funding USSOCOM Operations*, Santa Monica, Calif.: RAND Corporation, RR-360-SOCOM, 2014. As of January 17, 2020: https://www.rand.org/pubs/research_reports/RR360.html

Lyles, Ian Bradley, *Demystifying Counterinsurgency: U.S. Army Internal Security Training and South American Responses in the 1960s*, Austin: University of Texas, Austin, 2016.

Lynes, Jerome, "A Critique of 'Special Operations Doctrine: Is It Needed,'" *Prism*, Vol. 6, No. 4, 2017. As of November 13, 2018: https://cco.ndu.edu/News/Article/1171896/a-letters-to-the-editor-a-critique-of-special-operations-doctrine-is-it-needed/

MacKinnon, Mark, "In Bosnia-Herzegovina, Fears Are Growing That the Carefully Constructed Peace Is Starting to Unravel," *Globe and Mail*, May 24, 2018. As of February 14, 2020:
https://www.theglobeandmail.com/world/
article-in-bosnia-serb-nationalists-see-putin-and-trump-as-their-tickets-to/

Mallory, King, *New Challenges in Cross-Domain Deterrence*, Santa Monica, Calif.: RAND Corporation, PE-259-OSD, 2018. As of January 17, 2020:
https://www.rand.org/pubs/perspectives/PE259.html

Manwaring, Max G., and Court Prisk, *El Salvador at War: An Oral History from the 1979 Insurrection to the Present*, Washington, D.C.: National Defense University Press, 1988.

McRaven, William H., *Spec Ops: Case Studies in Special Warfare Operations: Theory and Practice*, New York: Presidio Press, 1995.

Mendel, William W., "Illusive Victory: From Blast Furnace to Green Sweep," *Military Review*, December 1992, pp. 74–87.

Menzel, Sewall H., *Fire in the Andes: U. S. Foreign Policy and Cocaine Politics in Bolivia and Peru*, Lanham, Md.: University Press of America, 1997.

Metz, Steven, *Rethinking Insurgency*, Carlisle, Pa.: Strategic Studies Institute, U.S. Army War College, 2007.

Meyer, Edward C., "The Challenge of Change," *Army 1981–82 Green Book*, October 1981.

Meyer, Victoria, "Southern Star Shines Brightly in Chile," *Tip of the Spear*, November 2009, pp. 16–21. As of February 15, 2020:
https://www.socom.mil/TipOfTheSpear/
November%202009%20Tip%20of%20the%20Spear.pdf

Miakhel, Shahmahmood, and Noah Coburn, *Many Shuras Do Not a Government Make: International Community Engagement with Local Councils in Afghanistan*, Washington, D.C.: U.S. Institute of Peace, September 7, 2010.

Miller, Paul D., "Obama's Failed Legacy in Afghanistan," *American Interest*, February 15, 2016. As of January 17, 2020:
https://www.the-american-interest.com/2016/02/15/
obamas-failed-legacy-in-afghanistan/

Mitchell, Eli G., *Special Operations Professional Military Education for Field Grade Officers*, Maxwell Air Force Base, Ala.: Air Command and Staff College, 2015.

Montgomery, Tommie Sue, *Revolution in El Salvador: Origins and Evolution*, Boulder, Colo.: Westview Press, 1982.

Moyar, Mark, Hector Pagan, and Wil R. Griego, *Persistent Engagement in Colombia*, Tampa, Fla.: Joint Special Operations University, July 2014.

National Security Decision Directive 221, *Narcotics and National Security*, Washington, D.C.: White House, April 8, 1986. As of February 2, 2018: https://fas.org/irp/offdocs/nsdd/nsdd-221.pdf

NATO—*See* North Atlantic Treaty Organization.

Newsom, Rob, "Adapting for the 'Other' War," *Small Wars Journal*, October 18, 2013.

North Atlantic Treaty Organization Stabilisation Force, "History of the NATO-Led Stabilisation Force (SFOR) in Bosnia and Herzegovina," undated. As of February 1, 2020: https://www.nato.int/sfor/docu/d981116a.htm

Odierno, Raymond T., James F. Amos, and William H. McRaven, *Strategic Landpower: Winning the Clash of Wills*, Washington, D.C.: U.S. Army, U.S. Marine Corps, and U.S. Special Operations Command, May 2013.

O'Hanlon, Michael, *The State of U.S. Military Readiness*, Washington, D.C.: Brookings Institution, August 15, 2016.

Oland, Dwight D., and David W. Hogan Jr., *Department of the Army Historical Summary: Fiscal Year 1992*, Washington, D.C.: U.S. Army Center of Military History, 2001.

O'Mahony, Angela, Thomas S. Szayna, Christopher G. Pernin, Laurinda L. Rohn, Derek Eaton, Elizabeth Bodine-Baron, Joshua Mendelsohn, Osonde A. Osoba, Sherry Oehler, Katharina Ley Best, and Leila Bighash, *The Global Landpower Network: Recommendations for Strengthening Army Engagement*, Santa Monica, Calif.: RAND Corporation, RR-1813-A, 2017. As of May 7, 2020: https://www.rand.org/pubs/research_reports/RR1813.html

O'Neal, Gary, and David Fisher, *American Warrior*, New York: Thomas Dunne Books, 2013.

Oreskes, Michael, "Poll Finds U.S. Expects Peace Dividend," *New York Times*, January 25, 1990.

Organization for Security and Co-operation in Europe, "Dayton Peace Agreement," December 14, 1995. As of January 7, 2019: https://www.osce.org/bih/126173

O'Shaughnessy, Terrence J., Matthew D. Strohmeyer, and Christopher D. Forrest, "Strategic Shaping: Expanding the Competitive Space," *Joint Forces Quarterly*, Vol. 90, No. 3, July 2018.

Paddock, Alfred, *U.S. Army Special Forces: Its Origins*, Washington, D.C.: National Defense University Press, 1982.

Paret, Peter, "Napoleon and the Revolution in War," in Peter Paret, Gordon A. Craig, and Felix Gilbert, eds., *Makers of Modern Strategy from Machiavelli to the Nuclear Age*, Princeton, N.J.: Princeton University Press, 1986.

Paul, Chris, Colin P. Clarke, Beth Grill, and Molly Dunigan, *Paths to Victory: Lessons from Modern Insurgencies*, Santa Monica, Calif.: RAND Corporation, RR-291/1-OSD, 2013. As of February 5, 2018: https://www.rand.org/pubs/research_reports/RR291z1.html

Peceny, Mark, and William D. Stanley, "Counterinsurgency in El Salvador," *Politics and Society*, Vol. 38, No. 1, 2010.

Pellerin, Cheryl, "Lynn: Cyberspace Is the New Domain of Warfare," U.S. Department of Defense, October 18, 2010. As of February 10, 2020: https://archive.defense.gov/news/newsarticle.aspx?id=61310

Peltier, Isaac J., *Surrogate Warfare: The Role of U.S. Army Special Forces*, Fort Leavenworth, Kan.: School of Advanced Military Studies, U.S. Army Command and General Staff College, 2005.

Pettersson, Thérése, and Kristine Eck, "Organized Violence, 1989–2017," *Journal of Peace Research*, Vol. 55, No. 4, 2018.

Phillips, R. Cody, *Operation Just Cause: The Incursion into Panama*, Washington, D.C.: U.S. Army Center for Military History, 2004. As of February 10, 2020: https://history.army.mil/html/books/070/70-85-1/cmhPub_70-85-1.pdf

Pike, Douglas, "Conduct of the War: Strategic Factors," in John Schlight, ed., *The Second Indochina War: Proceedings of a Symposium Held at Airlie, Virginia, 7–9 November 1984*, Washington, D.C.: U.S. Army Center of Military History, 1986.

Pinkston, Bobby Ray, *The Military Instrument of Power in Small Wars*, Fort Leavenworth, Kan.: U.S. Command and General Staff College, 1996.

Priest, Dana, "Covert Action in Colombia," *Washington Post*, December 21, 2013.

Public Law 97-86, Department of Defense Authorization Act, 1982, December 1, 1981.

Public Law 99-433, Goldwater-Nichols Department of Defense Reorganization Act of 1986, October 1, 1986.

Public Law 99-661, National Defense Authorization Act for Fiscal Year 1987, November 14, 1986.

Public Law 101-510, National Defense Authorization Act for Fiscal Year 1991, November 5, 1990.

Public Law 108-375, National Defense Authorization Act of Fiscal Year 2005, October 28, 2004.

Public Law 108-458, Intelligence Reform and Terrorism Prevention Act of 2004, December 17, 2004.

Public Law 112-25, Budget Control Act of 2011, August 2, 2011. As of May 21, 2020: https://www.govinfo.gov/app/details/PLAW-112publ25

Public Law 114-328, National Defense Authorization Act for Fiscal Year 2017, December 23, 2017. As of February 16, 2019:
https://www.govinfo.gov/content/pkg/PLAW-114publ328/html/PLAW-114publ328.htm

Public Law 115-91, National Defense Authorization Act for Fiscal Year 2018, December 12, 2017. As of May 7, 2020:
https://www.govinfo.gov/content/pkg/PLAW-115publ91/html/PLAW-115publ91.htm

Ramirez, Armando J., *From Bosnia to Baghdad: The Evolution of US Army Special Forces from 1995–2004*, Monterey, Calif.: Naval Postgraduate School, 2004.

Ramos, Alexis, "Joint, Combined Forces Conclude Southern Star," U.S. Southern Command, September 7, 2018. As of February 15, 2020:
https://www.southcom.mil/MEDIA/NEWS-ARTICLES/Article/1627031/joint-combined-forces-conclude-southern-star/

Reagan, Ronald, "Excerpts from an Interview with Walter Cronkite of CBS News," March 3, 1981. As of February 13, 2020:
https://www.reaganlibrary.gov/research/speeches/30381c

———, "Remarks on Signing Executive Order 12368, Concerning Federal Drug Abuse Policy Functions," June 24, 1982. As of February 3, 2018:
https://www.reaganlibrary.gov/research/speeches/62482b

Record, Jeffrey, *The American Way of War Cultural Barriers to Successful Counterinsurgency*, Washington, D.C.: CATO Institute, 2006.

Redden, Mark E., and Michael P. Hughes, "Defense Planning Paradigms and the Global Commons," *Joint Forces Quarterly*, Vol. 60, No. 1, 2011.

Rempe, Dennis M., *The Past as Prologue? A History of U.S. Counterinsurgency Policy in Colombia, 1958–66*, Carlisle, Pa.: Strategic Studies Institute, U.S. Army War College, 2002.

Robinson, Linda, *Masters of Chaos: The Secret History of the Special Forces*, New York: PublicAffairs, 2001.

———, *One Hundred Victories: Special Ops and the Future of American Warfare*, New York: PublicAffairs, 2013.

Robinson, Linda, Todd C. Helmus, Raphael S. Cohen, Alireza Nader, Andrew Radin, Madeline Magnuson, and Katya Migacheva, *Modern Political Warfare: Current Practices and Possible Responses*, Santa Monica, Calif.: RAND Corporation, RR-1772-A, 2018. As of June 15, 2020:
https://www.rand.org/pubs/research_reports/RR1772.html

Robinson, Linda, Austin Long, Kimberly Jackson, and Rebeca Orrie, *Improving the Understanding of Special Operations: A Case History Analysis*, Santa Monica, Calif.: RAND Corporation, RR-2026-A, 2018. As of May 7, 2020:
https://www.rand.org/pubs/research_reports/RR2026.html

Rohde, Joy, *Armed with Expertise: The Militarization of American Social Research During the Cold War*, Ithaca, N.Y.: Cornell University Press, 2013.

Rosello, Victor M., "Lessons from El Salvador," *Parameters*, Winter 1993, pp. 100–108.

Rosenau, William, *"Irksome and Unpopular Duties": Pakistan's Frontier Corps, Local Security Forces, and Counterinsurgency*," Alexandria, Va.: CNA, May 2012.

Rothstein, Hy, "Less Is More: The Problematic Future of Irregular Warfare in an Era of Collapsing States," *Third World Quarterly*, Vol. 28, No. 2, 2007, pp. 275–294.

Rothstein, Hy, and Barton Whaley, *The Art and Science of Military Deception*, Norwood, Mass.: Artech House, 2013.

Rottman, Gordon L., *Mobile Strike Forces in Vietnam, 1966–70*, London: Osprey, 2013.

Sacolick, Bennet, and Wayne W. Grigsby Jr., "Special Operations/Conventional Forces Interdependence: A Critical Role in Prevent, Shape, Win," *Army Magazine*, June 2012.

Salmoni, Barak A., Jessica Hart, Renny McPherson, and Aidan Kirby Win, "Growing Strategic Leaders for Future Conflict," *Parameters*, September–October 2018.

Schanzer, Jonathan, "Ansar Al-Islam: Back in Iraq," *Middle East Quarterly*, Winter 2004, pp. 41–50.

Schifrin, Nick, and Habibullah Khan, "3 U.S. Special Forces Die in Pakistan Bombing," ABC News, February 3, 2010. As of November 1, 2018: https://abcnews.go.com/International/us-military-die-pakistan-bombing/story?id=9734681

Schmitt, Eric, and Thom Shanker, "Bush Administration Reviews Its Afghanistan Policy, Exposing Points of Contention," *New York Times,* September 22, 2008.

Schmitt, Paul J., *Special Operations Liaison Efforts: (SOLO) or Team Effort?* Carlisle, Pa.: U.S. Army War College, 2013.

Schwarz, Benjamin C., *American Counterinsurgency Doctrine and El Salvador: The Frustrations of Reform and the Illusions of Nation Building*, Santa Monica, Calif.: RAND Corporation, R-4042-USDP, 1991. As of May 7, 2020: https://www.rand.org/pubs/reports/R4042.html

Scobell, Andrew, David Lai, and Roy Kamphausen, *Chinese Lessons from Other Peoples' Wars*, Carlisle, Pa.: Strategic Studies Institute, U.S. Army War College, 2011.

Seck, Hope Hodge, "Marines' Alliance with Georgians Holds Clues to Future Missions," *Marine Corps Times*, December 11, 2014. As of January 31, 2020: https://www.marinecorpstimes.com/news/pentagon-congress/2014/12/11/ marines-alliance-with-georgians-holds-clues-to-future-missions/

Seelke, Clare Ribando, *Bolivia: In Brief*, Washington, D.C.: Congressional Research Service, 2014. As of February 5, 2018: https://fas.org/sgp/crs/row/R43473.pdf

Seelke, Clare Ribando, Liana Sun Wyler, June S. Beittel, and Mark P. Sullivan, *Latin America and the Caribbean: Illicit Drug Trafficking and U.S. Counterdrug Programs*, Washington, D.C.: Congressional Research Service, March 19, 2012.

Shanker, Thom, "U.S. Pilot's Remains Found in Iraq After 18 Years," *New York Times*, August 2, 2009.

Shannon, Elaine, *Desperados: Latin Drug Lords, U.S. Lawmen and the War America Can't Win*, New York: Viking, 1988.

Sheehan, Michael, "Comparative Counterinsurgency Strategies: Guatemala and El Salvador," *Conflict*, Vol. 9, No. 2, 1989.

Sheikh, Fawzia, "DOD: 30 U.S., U.K. Personnel to Mentor Pakistan's Frontier Corps," *Inside the Pentagon*, January 10, 2008.

Sheridan, Mary Beth, "For Help in Rebuilding Mosul, U.S. Turns to Its Former Foes," *Washington Post*, April 25, 2003.

Smith, Douglas I., *Army Aviation in Operation Just Cause*, Carlisle, Pa.: U.S. Army War College, April 15, 1992. As of February 4, 2020: https://apps.dtic.mil/dtic/tr/fulltext/u2/a251409.pdf

Smith, Rupert, *The Utility of Force: The Art of War in the Modern World*, New York: Knopf, 2007.

Snow, Shawn, "A Plan Colombia for Afghanistan," *Foreign Policy*, February 3, 2016. As of October 18, 2018: https://foreignpolicy.com/2016/02/03/a-plan-colombia-for-afghanistan/

Sosa, David, "Peace Colombia: The Success of U.S. Foreign Assistance in South America," U.S. Global Leadership Coalition, May 10, 2017. As of January 14, 2019: https://www.usglc.org/blog/ peace-colombia-the-success-of-u-s-foreign-assistance-in-south-america/

Special Presidential Task Force Relating to Narcotics, Marijuana and Dangerous Drugs, *Task Force Report: Narcotics, Marijuana and Dangerous Drugs, Findings and Recommendations*, June 6, 1969.

Spirtas, Michael, "Toward One Understanding of Multiple Domains," *RAND Blog*, May 2, 2018. As of November 5, 2018:
https://www.rand.org/blog/2018/05/toward-one-understanding-of-multiple-domains.html

Stahr, Elvis J., Jr., "Foreword," in John F. Kennedy, *Special Warfare*, Washington, D.C., Office of the Chief of Information, Department of the Army, 1962.

Stavridis, James G., "Statement of Admiral James G. Stavridis, United States Navy Commander, United States Southern Command," Washington, D.C.: U.S. House of Representatives, Committee on Appropriations Subcommittee on Defense, March 5, 2008.

Stejskal, James, *Special Forces Berlin: Clandestine Cold War Operations of the US Army's Elite 1956–1990*, Philadelphia, Pa.: Casemate Publishers, 2017.

Swatek, Bruce R., *Role of Special Forces Liaison Elements in Future Multinational Operations*, Fort Leavenworth, Kan.: U.S. Army Command and General Staff College, 2002.

Synovitz, Ron, "Pentagon Wants More Funding for Pakistan Frontier Corps," Radio Free Europe, November 20, 2007.

Taylor, Steven J., Robert Bogdan, and Marjorie DeVault, *Introduction to Qualitative Research Methods: A Guidebook and Resource*, 4th ed., Hoboken, N.J.: Wiley, 2015.

Themnér, Lotta, *UCDP/PRIO Armed Conflict Dataset Codebook*, Version 18.1, Uppsala, Sweden, and Oslo, Norway: Uppsala Conflict Data Program and Centre for the Study of Civil Wars, International Peace Research Institute, 2018. As of July 24, 2018:
http://ucdp.uu.se/downloads/ucdpprio/ucdp-prio-acd-181.pdf

Thompson, Garry L., *Army Downsizing Following World War I, World War II, Vietnam, and a Comparison to Recent Army Downsizing*, Fort Leavenworth, Kan.: U.S. Army Command and General Staff College, 2002.

Tisdel, Michael D., Ken D. Teske, and William C. Fleser, *Theater Special Operations Commands Realignment*, MacDill Air Force Base, Fla.: U.S. Special Operations Command, 2014. As of February 15, 2020:
https://ntrl.ntis.gov/NTRL/dashboard/searchResults/titleDetail/ADA607289.xhtml

TRADOC—*See* U.S. Army Training and Doctrine Command.

Turnley, Jessica, *Retaining a Precarious Value as Special Operations Go Mainstream*, MacDill Air Force Base, Fla.: Joint Special Operations University Press, 2008.

United Nations, "Letter Dated 14 October 1997 from the Secretary-General Addressed to the President of the Security Council," S/1997/794, October 14, 1997. As of January 7, 2019:
https://www.nato.int/ifor/un/u971014a.htm

United Nations Office on Drugs and Crime, "Cocaine," in *World Drug Report 2009*, Vienna, 2009, pp. 79–105. As of February 5, 2018:
https://www.unodc.org/documents/data-and-analysis/tocta/4.Cocaine.pdf

———, *Bolivia: Monitoreo de Cultivos de Coca 2016*, La Paz, Bolivia, July 2017. As of November 16, 2018:
https://www.unodc.org/documents/crop-monitoring/Bolivia/2016_Bolivia_Informe_Monitoreo_Coca.pdf

United States Code, Title 18, Section 1385, Use of Army and Air Force as Posse Comitatus.

U.S. Air Force, *Irregular Warfare Strategy*, Washington, D.C., 2013. As of October 12, 2018:
https://fas.org/irp/doddir/usaf/iw-strategy.pdf

U.S. Army, "Strategic Landpower Forum," August 4, 2014. As of January 22, 2020:
https://www.army.mil/standto/2014-08-04

U.S. Army, Europe, "Military Operations: The U.S. Army in Bosnia and Herzegovina," Army in Europe Pamphlet 525-100, October 7, 2003. As of January 7, 2019:
https://fas.org/irp/doddir/army/ae-pam-525-100.pdf

U.S. Army Special Operations Command, "10th SFG(A) History," webpage, undated-a. As of May 19, 2020:
https://www.soc.mil/USASFC/Groups/10th/history.html

———, "Assessing Revolution and Insurgent Strategies (ARIS) Studies," webpage, undated-b. As of February 13, 2020:
https://www.soc.mil/ARIS/ARIS.html

U.S. Army Special Operations Command, "ARSOF 2022," special issue, *Special Warfare*, April–June 2013.

———, "ARSOF 2022: Part II," special issue, *Special Warfare*, July–September 2014.

———, "ARSOF Next: A Return to First Principles," special issue, *Special Warfare*, April–June 2015.

———, "Spirit of America," *Special Warfare*, January–June 2016.

———, "ARSOF 2035," special issue, *Special Warfare*, 2017.

U.S. Army Training and Doctrine Command, *Joint Low-Intensity Conflict Project: Final Report*, Fort Monroe, Va., 1987. As of February 11, 2020:
https://apps.dtic.mil/dtic/tr/fulltext/u2/a185971.pdf

———, *U.S. Army Functional Concept for Engagement*, Fort Eustis, Va., U.S. Army Training and Doctrine Command Publication 525-8-5, February 24, 2014.

USASOC—*See* U.S. Army Special Operations Command.

U.S. Department of Defense, *Irregular Warfare (IW) Joint Operating Concept (JOC)*, Version 1.0, Washington, D.C., September 2007. As of February 14, 2020: https://fas.org/irp/doddir/dod/iw-joc.pdf

———, *Report on Progress Toward Security and Stability in Afghanistan*, Washington, D.C., June 2008. As of January 17, 2020: https://www.hsdl.org/?view&did=487165

———, *Summary of the 2018 National Defense Strategy of the United States of America: Sharpening the American Military's Competitive Edge*, Washington, D.C., January 19, 2018. As of October 8, 2018: https://dod.defense.gov/Portals/1/Documents/pubs/ 2018-National-Defense-Strategy-Summary.pdf

U.S. General Accountability Office, *Federal Drug Interdiction Efforts Need Strong Central Oversight*, Washington, D.C., June 13, 1983.

U.S. House Armed Services Committee, *Another Crossroads? Professional Military Education Two Decades After the Goldwater-Nichols Act and the Skelton Panel*, Washington, D.C., 2010.

———, *Institutionalizing Irregular Warfare Capabilities: Hearing Before the Subcommittee on Emerging Threats and Capabilities of the Committee on Armed Services, House of Representatives*, Washington, D.C.: Government Printing Office, 2012. As of October 12, 2018: https://www.gpo.gov/fdsys/pkg/CHRG-112hhrg71528/html/CHRG- 112hhrg71528.htm

U.S. House of Representatives, statement of Ronald F. Lauve, Senior Associate Director, General Government Division, *Hearing Before the Subcommittee on Crime, House Committee on the Judiciary on Narcotics Enforcement Policy*, Washington, D.C., 1981.

———, *Foreign Assistance and Related Programs Appropriations for 1988: Hearings Before a Subcommittee of the Committee of Appropriations*, Washington, D.C.: Government Printing Office, 1987.

U.S. House of Representatives, Committee on Foreign Affairs, *Report of a Staff Study Mission to Peru, Bolivia, Colombia, and Mexico, November 19 to December 18, 1988*, Washington, D.C.: Government Printing Office, February 1989.

U.S. House of Representatives, Committee on Government Operations, *Stopping the Flood of Cocaine with Operation Snowcap: Is It Working?* Washington, D.C.: Government Printing Office, 1990.

U.S. Joint Chiefs of Staff, *Officer Professional Military Education Policy (OPMEP)*, Washington, D.C., September 5, 2012. As of January 17, 2019: https://usacac.army.mil/sites/default/files/documents/cace/LREC/ 2011_CJCSI_1800.01D_Ch1_OPMEP.pdf

————, *Joint Concept for Human Aspects of Military Operations*, Washington, D.C., October 19, 2016.

————, *Joint Concept for Integrated Campaigning*, Washington, D.C., March 16, 2018.

U.S. Marine Corps, *Small Wars Manual*, Washington, D.C., 2014.

U.S. Senate, *Drugs, Law Enforcement and Foreign Policy: Hearings Before the Subcommittee on Terrorism, Narcotics, and International Communications and International Economic Policy, Trade, Oceans, and Environment of the Committee on Foreign Relations, United States Senate*, Washington, D.C.: Government Printing Office, 1988.

————, *Department of Defense Authorization for Appropriations for Fiscal Year 2010: Hearing Before the Committee on Armed Services, United States Senate*, Part 5, Washington, D.C.: Government Printing Office, 2009. As of October 12, 2018:
https://www.gpo.gov/fdsys/pkg/CHRG-111shrg52624/html/
CHRG-111shrg52624.htm

Vidino, Lorenzo, "How Chechnya Became a Breeding Ground for Terror," *Middle East Quarterly*, Vol. 12, No. 3, Summer 2005.

Weigley, Russell F., *History of the United States Army*, New York: Macmillan, 1967.

Weiner, Tim, "With Iraq's O.K., a U.S. Team Seeks War Pilot's Body," *New York Times*, December 14, 1995.

————, "Gulf War's First U.S. Casualty Leaves Lasting Trail of Mystery," *New York Times*, December 7, 1997.

White, D. Jonathan, *Doctrine for Special Forces in Stability and Support Operations*, Fort Leavenworth, Kan.: School of Advanced Military Studies, Command and General Staff College, 2000.

Whitty, David, *The Iraqi Counter Terrorism Service*, Washington, D.C.: Brookings Institution, 2016.

Woerner, Fred, *Report of the El Salvador Military Strategy Assistance Team*, Washington, D.C.: U.S. Department of Defense, 1981.

World Bank, World Development Indicators, data set, accessed 2018.

Yagoub, Mimi, "Challenging the Cocaine Figures, Part I: Bolivia," *InSight Crime*, November 16, 2016. As of November 16, 2018:
https://www.insightcrime.org/news/analysis/
challenging-the-cocaine-figures-part-1-bolivia/

About the Authors

Lieutenant General (Retired) **Charles T. Cleveland**, an international and defense researcher at RAND, retired as commanding general, U.S. Army Special Operations Command, in 2015. Cleveland spent more than 30 years of his 37-year career in the U.S. Army serving with special operations, including command of 10th Special Forces Group (Airborne), the Combined Joint Special Operations Task Force–North during Operation Iraqi Freedom, Special Operations Command South, and Special Operations Command Central. He has a B.S. from the U.S. Military Academy at West Point.

Daniel Egel is an economist at RAND who focuses on U.S. policy-making at the nexus of development and stability. His research uses both qualitative and quantitative methods to study the effectiveness of counterinsurgency, counterterrorism, foreign internal defense, humanitarian assistance, and stabilization efforts. He has a Ph.D. in economics from the University of California, Berkeley.